Universitext

Universitext is a series of textbooks that presents material from a wide variety of mathematical disciplines at master's level and beyond. The books, often well class-tested by their author, may have an informal, personal even experimental approach to their subject matter. Some of the most successful and established books in the series have evolved through several editions, always following the evolution of teaching curricula, into very polished texts.

Thus as research topics trickle down into graduate-level teaching, first textbooks written for new, cutting-edge courses may make their way into *Universitext*.

Michel Benaïm • Tobias Hurth

Markov Chains on Metric Spaces

A Short Course

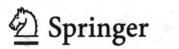 Springer

Michel Benaïm 🆔
Institut de mathématiques
Université de Neuchâtel
Neuchâtel, Switzerland

Tobias Hurth 🆔
Freie Universität Berlin
Berlin, Germany

This work was supported by Swiss National Science Foundation.

ISSN 0172-5939   ISSN 2191-6675   (electronic)
Universitext
ISBN 978-3-031-11821-0   ISBN 978-3-031-11822-7   (eBook)
https://doi.org/10.1007/978-3-031-11822-7

Mathematics Subject Classification: 60J05, 37-01, 37H15, 37B20, 60K15

This Springer imprint is published by the registered company Springer Nature Switzerland AG
The registered company address is: Gewerbestrasse 11, 6330 Cham, Switzerland

In memory of Marie Duflo

Preface

This book is based on a series of lectures given over recent years in Master's courses in probability. It provides a short, self-contained introduction to the ergodic theory of Markov chains on metric spaces.

Although primarily intended for graduate and postgraduate students, certain chapters (e.g., one and two) can be taught at the undergraduate level. Others (e.g., four and five) can be used as complements to courses in measure or ergodic theory. Basic knowledge in probability, measure theory, and calculus is recommended. A certain familiarity with discrete-time martingales is also useful, but the few results from martingale theory used in this book are all recalled in the appendix. Each chapter contains several exercises ranging from simple applications of the theory to more advanced developments and examples.

Whether in physics, engineering, biology, ecology, economics, or elsewhere, Markov chains are frequently used to describe the random evolution of complex systems. The understanding and analysis of these systems require, first of all, a good command of the mathematical techniques that allow to explain the long-term behavior of a general Markov chain living on a (reasonable) metric space. Presenting these techniques is, briefly put, our main objective. Questions that are central to this book and that will be recurrently visited are: under which conditions does such a chain have an invariant probability measure? If such a measure exists, is it unique? Does the empirical occupation measure of the chain converge? Does the law of the chain converge, and if so, in which sense and at which rate?

There are a variety of tools to address these questions. Some rely on purely measure-theoretic concepts that are natural generalizations of the ones developed for countable chains (i.e., chains living on countable state spaces). This includes notions of *irreducibility, recurrence* (in the sense of Harris), *petite and small sets*, etc. Other tools assume topological properties of the chain such as the *strong Feller* or *asymptotic strong Feller property* (in the sense of Hairer and Mattingly). However, when dealing with a specific model, measure-theoretic conditions—such as irreducibility—might be difficult to verify, and strong topological properties—such as the strong Feller condition—are seldom satisfied. A powerful approach is then to combine much weaker topological conditions—such as the (weak)

Feller condition—with controllability properties of the system to prove that certain measure-theoretic conditions (e.g., irreducibility, existence of petite or small sets) are satisfied. This approach is largely developed here and is a key feature of this book.

The book is organized in eight chapters and a short appendix. Chapter 1 briefly defines Markov chains and kernels and gives their very first properties, the Markov and strong Markov properties. The end of the chapter gives a concise introduction to Markov chains in continuous time, also called Markov processes, as they appear in many examples throughout the book.

Chapter 2 is a self-contained mini course on countable Markov chains. Classical notions of *recurrence (positive* and *null)* and *transience* are introduced. These are powerful notions, but when students meet them for the first time and have to verify that a specific chain is either recurrent or transient, they are often disoriented. Thus, we have chosen to spend some time here to show how these properties can be verified "in practice" with the help of suitable *Lyapunov functions*. We also explain how Lyapunov functions can be used to provide estimates on the moments (polynomial and exponential) of *hitting times* for a point or a finite set.

Certainly one of the most important results in the theory of countable chains is the *ergodic theorem*, which asserts that—for positive recurrent aperiodic chains—the law of the chain converges to a unique distribution. The final three sections of Chap. 2 are organized around this result. We first prove it quickly—by standard coupling—without any estimate on the rate of convergence. Then, the Lyapunov method is applied to investigate the behavior of *renewal processes* and provide short proofs of coupling theorems for these processes. Finally, relying on these coupling results, we revisit the ergodic theorem, this time with some convergence rates.

On uncountable state spaces, the simplest (and also the most natural) examples of Markov chains are given by *random dynamical systems* (also called *random iterative systems*). These are systems such that the state variable at time $n + 1$ is a deterministic function of the state variable at time n and a "random" input sampled from a sequence of i.i.d. random variables. Chapter 3 is devoted to this type of chain and explains how any given "abstract" Markov chain can be represented by a random dynamical system. Some interesting examples (Bernoulli convolutions, Propp-Wilson algorithm) are presented in exercises.

Chapter 4 starts with a detailed section on *weak convergence, tightness,* and *Prohorov's theorem*. Then, *invariant probability measures* are defined, and it is shown that, for a Feller chain, weak limit points for the family of *empirical occupation measures* are almost surely invariant probability measures. We discuss some practical tightness criteria (for the empirical occupation measures) based on Lyapunov functions. At this stage of the book, the reader understands that, under a reasonable control of the chain at infinity (obtained for instance by a Lyapunov function), uniqueness of the invariant probability measure equates stability: the empirical occupation measures converge almost surely to some (unique) distribution, regardless of the initial distribution. So we found it was a good place to discuss simple examples of *uniquely ergodic* chains (i.e., chains having a unique invariant probability measure). This is done in the third section of Chap. 4, where we analyze random dynamical systems obtained by random composition of contractions (or mappings that contract on average). The penultimate section of the chapter is devoted to ergodic theorems. We first prove several classical results (*Poincaré recurrence theorem, Birkhoff ergodic theorem,* and the *ergodic decomposition theorem*) and then show how they can be applied to Markov chains. Finally, we discuss invariant measures of continuous-time processes and explain how their properties (existence, ergodicity, uniqueness, ergodic decomposition, etc.) can be studied using discrete-time theory.

Chapter 5 is devoted to various notions of *irreducibility* which ensure unique ergodicity. We start with the measure-theoretic notion of irreducibility (also called ψ irreducibility) and then move on to more topological conditions. The *accessible set* of a Feller chain is introduced, and its relations with the support of invariant probability measures are investigated. We then consider strong Feller chains and prove that for such chains ergodic probability measures have disjoint support. We also prove the Hairer-Mattingly theorem, which says that the same property holds under the weaker assumption that the chain is *asymptotically strong Feller.* These results have the useful consequence that, on a connected set, if there is an invariant probability measure having full support, the chain is uniquely ergodic.

We then discuss in Chap. 6 the notions of *petite sets, small sets* and *(weak) Doeblin points* and show that the existence of an accessible weak Doeblin point implies irreducibility for (weak) Feller chains. This latter result is then applied to a variety of examples both in discrete time (random dynamical systems, random dynamical systems obtained by random switching between deterministic flows) and in continuous time (piecewise deterministic Markov processes, stochastic differential equations). This gives us the opportunity to show how the accessibility condition is naturally expressed as a control problem and how the Doeblin properties are naturally related to Hörmander type conditions (for random switching models, piecewise deterministic Markov processes, and SDEs).

Chapter 7 introduces *Harris recurrence.* For uniquely ergodic chains, Harris recurrence equates to *positive recurrence,* meaning that for every bounded Borel (and not merely for every continuous) function, the Birkhoff averages of the function converge almost surely. We prove the important result that Harris recurrence (respectively positive recurrence) is implied by the existence of a *recurrent petite set* (respectively a petite set whose first return time is bounded in L^1). We also discuss simple useful criteria (relying on Lyapunov functions) ensuring that a set is recurrent and provide moment estimates on the return times.

Chapter 8 revolves around the celebrated Harris ergodic theorem. After revisiting the notions of *total variation distance* and *coupling* for two probability measures, we state a simple version of the Harris ergodic theorem where the entire state space is a petite set. Under this strong hypothesis, one has exponential convergence in total variation distance to the unique invariant probability measure. The same conclusion holds under the existence of a Lyapunov function that forces the Markov chain to enter a certain small set—a condition that is better adapted to noncompact state spaces, which are usually not petite. We give two different proofs for this latter version of Harris's ergodic theorem: first the recent proof by Hairer and Mattingly based on the ingenious construction of a semi-norm for which the Markov operator is a contraction. And second, a more classical proof using coupling arguments and ideas from renewal theory. More precisely, under uniform estimates on polynomial (respectively exponential) moments for the return times to an aperiodic and recurrent small set, we obtain polynomial (respectively exponential) convergence in total variation distance to the unique invariant probability measure. Finally, we present a condition, also due to Hairer and Mattingly, that yields exponential convergence to the unique invariant probability measure in a certain Wasserstein distance.

The appendix recalls the monotone class theorem and the few results from discrete-time martingales that are used in the book.

More advanced textbooks include the excellent classical books by Meyn and Tweedie [49] and Duflo [22] and the more recent book by Douc, Moulines, Priouret, and Soulier [20]. The lecture notes by Hairer [31] contain some similar material and are also highly recommended.

Acknowledgments We first thank the Swiss National Science Foundation for its constant support over the years. This book has been prepared while we were working on the projects 200021-175728 and 200020-196999. We are grateful to Yuri Bakhtin, Jonathan Mattingly, and Tom Mountford for their valuable remarks. This book has also indirectly benefited from the numerous discussions we have had during its preparation with our colleagues, students, and post-docs: Jean-Baptiste Bardet, Antoine Bourquin, Charles-Edouard Bréhier, Nicolas Champagnat, Bertrand Cloez, Carl-Eric Gauthier, Eva Löcherbach, Florent Malrieu, Laurent Miclo, Pierre Monmarché, William Oçafrain, Fabien Panloup, Edouard Strickler, Sebastian Schreiber, Oliver Tough, Denis Villemonais, Pierre-André Zitt, and Alexandre Zumbrunnen.

Neuchatel, Switzerland   Michel Benaïm
Berlin, Germany   Tobias Hurth

Contents

Preliminaries

The general setting is the following. Throughout all this book, we let M denote a *separable* (there exists a countable dense subset) metric space with metric d (e.g., $\mathbb{R}$, $\mathbb{R}^n$) equipped with its Borel σ-field $\mathcal{B}(M)$. We let $B(M)$ (respectively $C_b(M)$) denote the set of real-valued bounded measurable (respectively bounded continuous) functions on M equipped with the norm

$$\|f\|_\infty := \sup_{x \in M} |f(x)|. \tag{1}$$

If μ is a (nonnegative) measure on M and $f \in L^1(\mu)$ (or $f \geq 0$ measurable), we let

$$\mu f := \int_M f(x)\, \mu(dx)$$

denote the integral of f with respect to μ. The rest of the notation is introduced in the main body of the text. Please also refer to the list of symbols at the end of the book.

Chapter 1
Markov Chains

This chapter introduces the basic objects of the book: Markov kernels and Markov chains. The Chapman-Kolmogorov equation, which characterizes the evolution of the law of a Markov chain, as well as the Markov and strong Markov properties are established. The last section briefly defines continuous-time Markov processes.

1.1 Markov Kernels

A *Markov kernel* on M is a family of measures

$$P = \{P(x, \cdot)\}_{x \in M}$$

such that

(i) For all $x \in M$, $P(x, \cdot) : \mathcal{B}(M) \to [0, 1]$ is a probability measure;
(ii) For all $G \in \mathcal{B}(M)$, the mapping $x \in M \mapsto P(x, G) \in \mathbb{R}$ is measurable.

The Markov kernel P acts on functions $g \in B(M)$ and measures (respectively probability measures) according to the formulae:

$$Pg(x) := \int_M P(x, dy)g(y), \tag{1.1}$$

$$\mu P(G) := \int_M \mu(dx)P(x, G). \tag{1.2}$$

M. Benaïm, T. Hurth, *Markov Chains on Metric Spaces*, Universitext,
https://doi.org/10.1007/978-3-031-11822-7_1

Remark 1.1 For all $g \in B(M)$, we have $Pg \in B(M)$ and $\|Pg\|_\infty \leq \|g\|_\infty$. Boundedness is immediate and measurability easily follows from the condition **(ii)** defining a Markov kernel (use for example the monotone class theorem from the appendix).

Remark 1.2 The term $Pg(x)$ can also be defined by (1.1) for measurable functions $g : M \to \mathbb{R}$ that are nonnegative, but not necessarily bounded. For such g, $Pg(x)$ is an element of $[0, \infty]$. This will play a role in the study of Lyapunov functions starting in Sect. 2.3.

We let P^n denote the operator recursively defined by $P^0 g := g$ and $P^{n+1} g := P(P^n g)$ for $n \in \mathbb{N}$. Or, equivalently,

$$P^0(x, \cdot) := \delta_x \text{ and } P^{n+1}(x, G) := \int_M P^n(x, dy) P(y, G)$$

for all $n \in \mathbb{N}$ and for all $G \in \mathcal{B}(M)$. Here and throughout these notes, $\mathbb{N}$ is the set of nonnegative integers (including 0). The set of positive integers (excluding 0) will be denoted by $\mathbb{N}^*$.

Example 1.3 (Countable Space) Suppose M is countable. We can turn M into a separable (and complete) metric space by endowing it with the discrete metric $d(x, y) = \mathbf{1}_{x \neq y}$. The corresponding Borel σ-field is the collection of all subsets of M. A *Markov transition matrix* on M is a map $P : M \times M \to [0, 1]$ such that

$$\sum_{y \in M} P(x, y) = 1$$

for all $x \in M$. This gives rise to a Markov kernel Q defined by

$$Q(x, G) := \sum_{y \in G} P(x, y)$$

for all $G \subset M$. Since there is a one-to-one correspondence between transition matrices and kernels on M, we shall identify P with Q and refer to it at times as a transition matrix and at times as a kernel.

1.2 Markov Chains

In order to define Markov chains, we first need to introduce the (classical) notions of filtration and adapted processes. Let $(\Omega, \mathcal{F}, \mathsf{P})$ be a probability space. A *filtration* $\mathbb{F} = (\mathcal{F}_n)_{n \geq 0}$ is an increasing sequence of σ-fields: $\mathcal{F}_n \subset \mathcal{F}_{n+1} \subset \mathcal{F}$ for all $n \in \mathbb{N}$. The data $(\Omega, \mathcal{F}, \mathbb{F}, \mathsf{P})$ is called a *filtered* probability space. An M-valued *adapted* stochastic process on $(\Omega, \mathcal{F}, \mathbb{F}, \mathsf{P})$ is a family $(X_n)_{n \geq 0}$ of random variables defined on $(\Omega, \mathcal{F}, \mathsf{P})$, taking values in M and such that X_n is $\mathcal{F}_n$-measurable for all $n \in \mathbb{N}$. If $X = (X_n)_{n \geq 0}$ is a family of random variables on $(\Omega, \mathcal{F}, \mathsf{P})$, the *canonical filtration* of X is the filtration $\mathbb{F}^X = \{\mathcal{F}_n^X\}_{n \geq 0}$ where $\mathcal{F}_n^X = \sigma(X_0, \ldots, X_n)$ is the σ-field generated by $X_0, \ldots, X_n$. With such a definition X is always an *adapted* stochastic process on $(\Omega, \mathcal{F}, \mathbb{F}^X, \mathsf{P})$.

We can now define what a Markov chain is. Given a filtered probability space $(\Omega, \mathcal{F}, \mathbb{F}, \mathsf{P})$ and a Markov kernel P on M, a *Markov chain with kernel P with respect to* $\mathbb{F}$ is an M-valued adapted stochastic process (X_n) on $(\Omega, \mathcal{F}, \mathbb{F}, \mathsf{P})$ such that

$$\mathsf{P}(X_{n+1} \in G | \mathcal{F}_n) = P(X_n, G)$$

for all $n \in \mathbb{N}$ and for all $G \in \mathcal{B}(M)$. Equivalently,

$$\mathsf{E}(g(X_{n+1}) | \mathcal{F}_n) = Pg(X_n)$$

for all $n \in \mathbb{N}$ and for all $g \in B(M)$ (or all functions $g : M \to \mathbb{R}$ that are measurable and nonnegative). Here, $\mathsf{E}(\cdot | \mathcal{F}_n)$ denotes conditional expectation with respect to $\mathcal{F}_n$, and $\mathsf{P}(X_{n+1} \in G | \mathcal{F}_n) := \mathsf{E}(1_{X_{n+1} \in G} | \mathcal{F}_n)$. In the appendix, we recall the definition of conditional expectation and list some of its basic properties, which will be used without further comment throughout the text.

Proposition 1.4 *Let (X_n) be a Markov chain with kernel P with respect to $\mathbb{F}$. Then (X_n) is always a Markov chain with kernel P with respect to $\mathbb{F}^X$. This latter property is equivalent to*

$$\mathsf{E}(g(X_{n+1})h_0(X_0) \ldots h_n(X_n)) = \mathsf{E}(Pg(X_n)h_0(X_0) \ldots h_n(X_n))$$

for all $n \in \mathbb{N}$, $h_0, \ldots, h_n \in B(M)$, and $g \in B(M)$.

Proof Suppose that (X_n) is a Markov chain with kernel P with respect to $\mathbb{F}$. Since $\mathcal{F}_n^X \subset \mathcal{F}_n$,

$$\mathsf{E}(g(X_{n+1}) | \mathcal{F}_n^X) = \mathsf{E}(\mathsf{E}(g(X_{n+1}) | \mathcal{F}_n) | \mathcal{F}_n^X) = Pg(X_n).$$

This proves the first statement. Multiplying the left-hand side and right-hand side of this equality by $h_0(X_0) \ldots h_n(X_n)$ and taking expected value shows the forward implication of the second statement. The backward implication follows from the definition of conditional expectation. $\square$

Remark 1.5 In view of Proposition 1.4, when we say that (X_n) is a Markov chain with kernel P, we implicitly mean that it is a Markov chain with respect to $\mathbb{F}^X$.

Given a Markov kernel P and a probability measure ν on M, there always exists a Markov chain (X_n) with kernel P and such that X_0 has law ν. As outlined in Proposition 1.8 (ii), this follows from the Ionescu-Tulcea theorem.

Proposition 1.6 (Chapman-Kolmogorov Equation) *Let (X_n) be a Markov chain with kernel P. Let μ_n denote the law of X_n. Then, for every $n \in \mathbb{N}$,*

$$\mu_{n+1} = \mu_n P = \mu_0 P^{n+1}.$$

Proof For every $g \in B(M)$,

$$\mu_{n+1} g = \mathsf{E}(g(X_{n+1})) = \mathsf{E}(\mathsf{E}(g(X_{n+1})|\mathcal{F}_n)) = \mathsf{E}(Pg(X_n)) = \mu_n Pg.$$

$\square$

Example 1.7 (Countable Space) Let (X_n) be a Markov chain on a countable state space M, with transition matrix P and initial distribution μ_0. The law μ_n of the random variable X_n then satisfies

$$\mu_n(\{x\}) = \sum_{y \in M} \mu_0(\{y\}) P^n(y, x), \quad \forall x \in M,$$

where P^n is the nth power of the matrix P. In matrix-vector notation, this identity can be written as

$$\mu_n = \mu_0 P^n,$$

where μ_n and μ_0 are row vectors. In particular, if μ_0 is the Dirac measure at a point $y \in M$, then the law of X_n assigns mass $P^n(y, x)$ to every singleton $\{x\}$, i.e.,

$$\mathsf{P}(X_n = x | X_0 = y) = P^n(y, x).$$

Feller and Strong Feller Chains
The Markov kernel P (or the associated Markov chain (X_n)) is said to be *Feller* if it takes bounded continuous functions into bounded continuous functions. It is said to be *strong Feller* if it takes bounded Borel functions into bounded continuous functions. If M is countable and equipped with the discrete metric, then every function on M is continuous. In particular, every Markov kernel on a countable set is strong Feller.

1.3 The Canonical Chain

Let $X = (X_n)_{n \geq 0}$ be a Markov chain with kernel P. Then X can be seen as a random variable on $(\Omega, \mathcal{F}, \mathsf{P})$ taking values in the *space of trajectories*

$$M^{\mathbb{N}} := \{\mathbf{x} = (x_i)_{i \in \mathbb{N}} : x_i \in M\}$$

equipped with the product σ-field $\mathcal{B}(M)^{\otimes \mathbb{N}}$ (see Exercise 1.9).

If X_0 has law ν, we let $\mathbb{P}_\nu$ denote the law of X, which is the image measure of P by X. In particular, for all Borel sets $A_0, \ldots, A_k \subset M$,

$$\mathsf{P}(X_0 \in A_0, \ldots, X_k \in A_k) = \mathbb{P}_\nu \{\mathbf{x} \in M^{\mathbb{N}} : (x_0, \ldots, x_k) \in A_0 \times \ldots \times A_k\}. \tag{1.3}$$

We let $\mathbb{E}_\nu$ denote the corresponding expectation. If ν is the Dirac measure at x, we use the standard notation $\mathbb{P}_x := \mathbb{P}_{\delta_x}$ and $\mathbb{E}_x := \mathbb{E}_{\delta_x}$.

Proposition 1.8

(i) Let $X = (X_n)_{n \geq 0}$ be a Markov chain with kernel P and initial distribution ν. Then, for all Borel sets $A_0, \ldots, A_k \subset M$,

$$\mathbb{P}_\nu \{\mathbf{x} \in M^{\mathbb{N}} : (x_0, \ldots, x_k) \in A_0 \times \ldots \times A_k\} =$$
$$\int_{A_0} \nu(dx_0) \int_{A_1} P(x_0, dx_1) \ldots \int_{A_k} P(x_{k-1}, dx_k). \tag{1.4}$$

(ii) Let $\Omega = M^{\mathbb{N}}$ and let $\mathcal{F} = \mathcal{B}(M)^{\otimes \mathbb{N}}$. Given a probability measure ν and a Markov kernel P on M, there exists a unique probability measure $\mathbb{P}_\nu$ on $(\Omega, \mathcal{F})$ characterized by (1.4). On $(\Omega, \mathcal{F})$, the process $(X_n)_{n \geq 0}$ defined by $X_n(\mathbf{x}) = x_n$, is a Markov chain with kernel P and initial law ν, called the canonical chain.

Proof Given $k \in \mathbb{N}$ and $h_0, \ldots, h_k \in B(M)$, we let $h_0 \otimes \ldots \otimes h_k$ denote the map on $M^{\mathbb{N}}$ defined as

$$h_0 \otimes \ldots \otimes h_k(\mathbf{x}) := h_0(x_0) \ldots h_k(x_k).$$

For further reference such a map will be called a *product map* of length $k + 1$. Then

$$\mathsf{E}(h_0(X_0) \ldots h_k(X_k)) = \mathbb{E}_\nu(h_0 \otimes \ldots \otimes h_k)$$
$$= \mathbb{E}_\nu(h_0 \otimes \ldots \otimes h_{k-1} P h_k) = \nu[h_0 P[h_1 P[\ldots h_{k-1} P h_k] \ldots]]. \tag{1.5}$$

The first equality is by definition of $\mathbb{E}_\nu$. The second equality follows from Proposition 1.4 and the last one follows from the second one by induction on k. This proves the first statement.

The existence of a unique probability measure $\mathbb{P}_\nu$ on $(\Omega, \mathcal{F})$ characterized by (1.4) is the celebrated Ionescu-Tulcea theorem (see, e.g., Theorem 2 in Chapter II.9 of [63]). Using the result from Exercise 1.9, it is not hard to check that the canonical process (X_n) is a Markov chain on the filtered probability space $(\Omega, \mathcal{F}, \mathbb{F}^X, \mathbb{P}_\nu)$, with initial distribution ν and kernel P.   □

Exercise 1.9 Let $\mathcal{B}(M^n)$ (respectively $\mathcal{B}(M^{\mathbb{N}})$) denote the Borel σ-field over M^n (respectively $M^{\mathbb{N}}$, endowed with the product topology). Let $\mathcal{B}(M)^{\otimes n}$ (respectively $\mathcal{B}(M)^{\otimes \mathbb{N}}$) denote the product σ-field over M^n (respectively $M^{\mathbb{N}}$). Show that $\mathcal{B}(M)^{\otimes n} = \mathcal{B}(M^n)$ and $\mathcal{B}(M)^{\otimes \mathbb{N}} = \mathcal{B}(M^{\mathbb{N}})$.

Hint: For the inclusion $\subset$ one can use the fact that the projection $\pi_i : M^{\mathbb{N}} \to M, \mathbf{x} \mapsto x_i$ is continuous, hence measurable. Observe that this doesn't require the separability of M. For the converse implication, one can first show, using separability, that every open subset of M^n is a countable union of product sets $O_1 \times \ldots \times O_n$ with O_i open.

1.4   Markov and Strong Markov Properties

For $n \in \mathbb{N}$, we let $\Theta^n : M^{\mathbb{N}} \to M^{\mathbb{N}}$ denote the *shift operator* defined by

$$\Theta^n(\mathbf{x}) := (x_{n+k})_{k \geq 0}.$$

The following proposition known as the *Markov property* easily follows from the definitions.

Proposition 1.10 (Markov Property) *Let* $H : M^{\mathbb{N}} \to \mathbb{R}$ *be a nonnegative or bounded measurable function and* X *a Markov chain with kernel* P. *Then*

$$\mathsf{E}(H(\Theta^n \circ X)|\mathcal{F}_n) = \mathbb{E}_{X_n}(H).$$

Proof Assume without loss of generality that H is bounded. Indeed, if H is nonnegative and unbounded, there is an increasing sequence of bounded nonnegative functions that converges pointwise to H, and one can apply the monotone convergence theorem. The set of bounded H satisfying the required property is a vector space, containing the constant functions and closed under bounded monotone convergence. Therefore, by the monotone class theorem (given in the appendix) and by Exercise 1.9, it suffices to check the property when $H = h_0 \otimes \ldots \otimes h_k$ is a product map. We proceed by induction on k. If $k = 0$, this is immediate. If the property holds for all product maps of length $k + 1$, then

$$\mathsf{E}(h_0(X_n)\ldots h_k(X_{n+k})h_{k+1}(X_{n+k+1})|\mathcal{F}_n)$$

$$= \mathsf{E}(h_0(X_n)\ldots h_k(X_{n+k})\mathsf{E}(h_{k+1}(X_{n+k+1})|\mathcal{F}_{n+k})|\mathcal{F}_n)$$

$$= \mathsf{E}(h_0(X_n)\ldots h_k(X_{n+k})Ph_{k+1}(X_{n+k})|\mathcal{F}_n) = \mathbb{E}_{X_n}(h_0 \otimes \ldots \otimes h_k Ph_{k+1}).$$

By (1.5), this last term equals $\mathbb{E}_{X_n}(h_0 \otimes \ldots \otimes h_{k+1})$. $\square$

A *stopping time* on a filtered probability space $(\Omega, \mathcal{F}, \mathbb{F}, \mathsf{P})$ is a random variable $T : \Omega \to \mathbb{N} \cup \{\infty\}$ such that for all $n \in \mathbb{N}$, the event $\{T = n\} = T^{-1}(\{n\})$ lies in $\mathcal{F}_n$. The *σ-field generated by T*, denoted $\mathcal{F}_T$, is the σ-field consisting of all events $A \in \mathcal{F}$ such that

$$A \cap \{T = n\} \in \mathcal{F}_n, \quad \forall n \in \mathbb{N}.$$

Exercise 1.11

(i) Show that $\mathcal{F}_T$ is indeed a σ-field.
(ii) Let $(T_n)_{n\in\mathbb{N}}$ be a sequence of stopping times on a filtered probability space $(\Omega, \mathcal{F}, \mathbb{F}, \mathsf{P})$ such that $T_n \leq T_{n+1}$ for every $n \in \mathbb{N}$. Show that $\mathcal{A}_n := \mathcal{F}_{T_n}$, $n \in \mathbb{N}$, defines a filtration on $(\Omega, \mathcal{F}, \mathsf{P})$.

The following proposition generalizes Proposition 1.10.

Proposition 1.12 (Strong Markov Property) *Let $H : M^{\mathbb{N}} \to \mathbb{R}$ be a nonnegative or bounded measurable function, X a Markov chain, and T a stopping time living on the same filtered probability space as X. Then*

$$\mathsf{E}(H(\Theta^T \circ X)|\mathcal{F}_T)\mathbf{1}_{T<\infty} = \mathbb{E}_{X_T}(H)\mathbf{1}_{T<\infty}.$$

Proof It suffices to show that for all $n \in \mathbb{N}$,

$$\mathsf{E}(H(\Theta^n \circ X)\mathbf{1}_{T=n}|\mathcal{F}_T) = \mathbb{E}_{X_n}(H)\mathbf{1}_{T=n}.$$

The right-hand side is $\mathcal{F}_T$-measurable, and for all $A \in \mathcal{F}_T$,

$$\mathsf{E}(H(\Theta^n \circ X)\mathbf{1}_{T=n}\mathbf{1}_A) = \mathsf{E}(\mathbb{E}_{X_n}(H)\mathbf{1}_{T=n}\mathbf{1}_A)$$

by the Markov property (because $\mathbf{1}_{T=n}\mathbf{1}_A$ is $\mathcal{F}_n$-measurable). This proves the result.
                                                                              $\square$

1.5   Continuous Time: Markov Processes

Although this book is about Markov chains in discrete time, it is useful to say a
few words about Markov chains in continuous time, also called Markov processes,
because they appear in many examples throughout the book. The definitions are
modeled on discrete time. A *Markov semigroup* on M is a family $\{P_t\}_{t\geq 0}$ of Markov
kernels on M such that

(i) $P_0(x, \cdot) = \delta_x$;
(ii) For all $G \in \mathcal{B}(M)$, the mapping $(t, x) \rightarrow P_t(x, G)$ is measurable;
(iii) For all $t, s \geq 0$, $P_{t+s} = P_t \circ P_s$.

Let $(\Omega, \mathcal{F}, \mathsf{P})$ be a probability space and let $\mathbb{F} = (\mathcal{F}_t)_{t\geq 0}$ be a *continuous-time
filtration*, i.e., a family of σ-fields such that $\mathcal{F}_s \subset \mathcal{F}_t \subset \mathcal{F}$ for all $0 \leq s \leq t$.
An M-valued *adapted* stochastic process on $(\Omega, \mathcal{F}, \mathbb{F}, \mathsf{P})$ is a family $(X_t)_{t\geq 0}$ of
random variables defined on $(\Omega, \mathcal{F}, \mathsf{P})$, taking values in M and such that X_t is $\mathcal{F}_t$-
measurable for all $t \geq 0$.

A *Markov process* with semigroup $\{P_t\}_{t\geq 0}$ with respect to $\mathbb{F}$ is an adapted
stochastic process $X = (X_t)_{t\geq 0}$ on $(\Omega, \mathcal{F}, \mathbb{F}, \mathsf{P})$ such that for all $g \in B(M)$ and
$t, s \geq 0$,

$$\mathsf{E}(g(X_{t+s})|\mathcal{F}_t) = (P_s g)(X_t).$$

Exercise 1.13 Suppose M is countable. Let (Y_n) be a Markov chain on M with
kernel P. Let $U_1, U_2, \ldots$ be a sequence of independent identically distributed
random variables on $(0, \infty)$ having an exponential distribution of parameter λ, i.e.,
$\mathsf{P}(U_i > t) = e^{-\lambda t}$. Set $T_0 = 0$ and $T_n = U_1 + \ldots + U_n$ for $n \geq 1$. Let $(X_t)_{t\geq 0}$
be the continuous-time process defined by $X_t = Y_n$ for $T_n \leq t < T_{n+1}$. Show that
(X_t) is a Markov process with semigroup

$$P_t = e^{-\lambda t} e^{\lambda t P} := e^{-\lambda t} \sum_{k\geq 0} \frac{(\lambda t)^k P^k}{k!}.$$

Feller Processes
We use the following terminology. We say that the Markov semigroup $\{P_t\}_{t\geq 0}$ is
weak Feller provided that

(i) $P_t(C_b(M)) \subset C_b(M)$ for all $t \geq 0$;
(ii) For all $f \in C_b(M)$ and $x \in M$, $\lim_{t\downarrow 0} P_t f(x) = f(x)$.

This definition implies that P_t is Feller for all $t \geq 0$. Observe however that it is weaker than the usual definition of a *Feller semigroup* (see, e.g., [26, 59] or [45]), which assumes that

  (i) M is a locally compact metric space;
  (ii) $\{P_t\}_{t\geq 0}$ is a strongly continuous semigroup on $C_0(M)$ (the set of continuous functions vanishing at infinity), meaning that

  (a) $P_t(C_0(M)) \subset C_0(M)$;
  (b) For all $f \in C_0(M)$, $\lim_{t\downarrow 0} \|P_t f - f\|_\infty = 0$.

Remark 1.14 It is proved in [59, Proposition 2.4] that [(a), (b)] above is equivalent to [(a), (b)'] where (b)' is given by the (seemingly) weaker condition that

$$\lim_{t\downarrow 0} P_t f(x) = f(x)$$

for all $f \in C_0(M)$ and $x \in M$. As shown by the following exercise, this equivalence does not hold if $C_0(M)$ is replaced by $C_b(M)$.

Exercise 1.15 Let $M = (0, \infty)$, and let P_t be defined on $B(M)$ as

$$P_t f(x) = f\left(\frac{xe^t}{1 + x(e^t - 1)}\right).$$

Show that $\{P_t\}_{t\geq 0}$ is a weak Feller Markov semigroup which is not Feller.

Chapter 2
Countable Markov Chains

This chapter presents the basic theory of countable Markov chains. The assumption that M is countable makes the proofs easier and permits to introduce, in a simple setting, some of the key notions (such as *invariant probability measures, irreducibility, positive recurrence, etc.*) that will be revisited in the subsequent chapters. Furthermore, some of the results given here, in particular in Sect. 2.6, will be used later to prove the main results in Chap. 7. We assume here that M is a countable set equipped with the σ-field $\mathcal{S}$ of all subsets of M, and (X_n) is a Markov chain on M with Markov kernel (or matrix) $P = P(x, y)_{x,y \in M}$. In most of this chapter, we assume without loss of generality that $\Omega = M^{\mathbb{N}}, \mathcal{F} = \mathcal{S}^{\otimes \mathbb{N}}, X_n(\omega) = \omega_n$, and $\mathcal{F}_n = \sigma(X_0, \ldots, X_n)$, i.e., (X_n) is the canonical chain introduced in Sect. 1.3.

2.1 Recurrence and Transience

For $x \in M$, we let

$$\tau_x := \inf\{k \geq 1 : X_k = x\}$$

denote the first time ≥ 1 at which the chain hits x,

$$\tau_x^{(n)} := \inf\{k > \tau_x^{(n-1)} : X_k = x\},$$

the nth time of hitting x (with $\tau_x^{(0)} := 0$), and

$$N_x := \sum_{k \geq 1} \mathbf{1}_{\{X_k = x\}} \in \mathbb{N} \cup \{\infty\}$$

© The Author(s), under exclusive license to Springer Nature Switzerland AG 2022
M. Benaïm, T. Hurth, *Markov Chains on Metric Spaces*, Universitext,
https://doi.org/10.1007/978-3-031-11822-7_2

the number of visits of x at or after time 1. We adopt the convention that $\inf \emptyset = +\infty$. A point x is said to be *recurrent* if

$$\mathbb{P}_x(\tau_x < \infty) = 1$$

and *transient* otherwise.

Given $x, y \in M$ and $k \in \mathbb{N}^*$, we say that x *leads to* y *in k steps*, written $x \leadsto^k y$, if $P^k(x, y) > 0$. We say that x *leads to* y, written $x \leadsto y$, if $x \leadsto^k y$ for some $k \in \mathbb{N}^*$. The chain is called *irreducible* if $x \leadsto y$ for all $x, y \in M$. To any Markov chain on a countable set M with transition matrix P, one can associate a weighted directed graph as follows: Let M be the set of vertices. For any $x, y \in M$, not necessarily distinct, there is a directed edge of weight $P(x, y)$ going from x to y if and only if $P(x, y) > 0$. The chain is then irreducible if and only if the associated directed graph is connected, i.e., for any $x, y \in M$ there is a path from vertex x to vertex y that moves along directed edges. Note that a general notion of irreducibility will be defined in Chap. 5 and that every countable irreducible chain (as defined here) satisfies this general definition.

Proposition 2.1

(i) *If x is transient, then $N_x < \infty$ a.s. and for all $k \geq 0$,*

$$\mathbb{P}_x(N_x = k) = a^k(1 - a),$$

where $a = \mathbb{P}_x(\tau_x < \infty)$. In particular,

$$\mathbb{E}_x(N_x) = \sum_{k \geq 1} P^k(x, x) = \frac{a}{1 - a} < \infty.$$

(ii) *If x is recurrent, then $\mathbb{P}_x(N_x = \infty) = 1$,*

$$\mathbb{E}_x(N_x) = \sum_{k \geq 1} P^k(x, x) = \infty,$$

and

$$\lim_{n \to \infty} \frac{1}{n} \sum_{k=1}^{n} \mathbf{1}_{\{X_k = x\}} = \frac{1}{\mathbb{E}_x(\tau_x)}$$

$\mathbb{P}_x$-*a.s.*

(iii) *If the chain is irreducible, then either all points are recurrent or all points are transient. In the recurrent case, for all $x, y \in M$,*

$$\mathbb{P}_x(\tau_y < \infty) = 1 \text{ and } \mathbb{E}_x(N_y) = \infty.$$

In the transient case, for all $x, y \in M$,

$$\mathbb{E}_x(N_y) < \infty.$$

Proof

(i) Using the strong Markov property,

$$\mathbb{P}_x(N_x = k) = \mathbb{P}_x(\tau_x^{(k)} < \infty; \tau_x^{(k+1)} = \infty) = (1-a)\mathbb{P}_x(\tau_x^{(k)} < \infty)$$

and

$$\mathbb{P}_x(\tau_x^{(k)} < \infty) = a\mathbb{P}_x(\tau_x^{(k-1)} < \infty) = \ldots = a^k.$$

(ii) If x is recurrent, then, using again the strong Markov property,

$$\mathbb{P}_x(\tau_x^{(n)} < \infty) = \mathbb{P}_x(\tau_x^{(n-1)} < \infty) = \ldots = 1.$$

Hence $\mathbb{P}_x(N_x = \infty) = 1$ and thus $\mathbb{E}_x(N_x) = \infty$.

For all $n \geq 1$, there exists $k(n) \geq 0$ such that $\tau_x^{(k(n))} \leq n < \tau_x^{(k(n)+1)}$. Furthermore, the random variables $(\tau_x^{(n+1)} - \tau_x^{(n)})_{n \geq 0}$ are, under $\mathbb{P}_x$, i.i.d. Thus, by the strong law of large numbers for nonnegative i.i.d. random variables,

$$\lim_{n \to \infty} \frac{1}{n} \sum_{k=1}^{n} \mathbf{1}_{\{X_k = x\}} = \lim_{n \to \infty} \frac{k(n)}{\tau_x^{(k(n))}} = \frac{1}{\mathbb{E}_x(\tau_x)}.$$

(iii) If the chain is irreducible, for all $x, y \in M$ there exist $i, j \geq 1$ and $\varepsilon > 0$ such that $P^i(x, y) \geq \varepsilon$, $P^j(y, x) \geq \varepsilon$. Thus $P^{k+i+j}(x, x) \geq \varepsilon^2 P^k(y, y)$ for all $k \geq 1$. Therefore, we have the implication

$$\sum_{k \geq 1} P^k(y, y) = \infty \quad \Rightarrow \quad \sum_{k \geq 1} P^k(x, x) = \infty,$$

proving that x is recurrent whenever y is recurrent and y is transient whenever x is transient.

Suppose the chain is recurrent. Fix $x, y \in M$ such that $x \neq y$ (for $x = y$ the statement holds trivially true). By irreducibility, recurrence, and the strong Markov property,

$$\varepsilon := \mathbb{P}_x(\exists k < \tau_x : X_k = y) > 0.$$

Thus, again using the strong Markov property,

$$\mathbb{P}_x(\tau_y > \tau_x^{(n+1)}) = \mathbb{E}_x(\mathbb{P}_x(\tau_y > \tau_x^{(n+1)}|\mathcal{F}_{\tau_x^{(n)}}))$$

$$= \mathbb{E}_x((1 - \mathbb{P}_x(\exists k < \tau_x : X_k = y))\mathbf{1}_{\tau_y > \tau_x^{(n)}})$$

$$= (1 - \varepsilon)\mathbb{P}_x(\tau_y > \tau_x^{(n)}) = \ldots = (1 - \varepsilon)^{n+1}.$$

Thus $\mathbb{P}_x(\tau_y > \tau_x^{(n+1)}) \to 0$ as $n \to \infty$, showing that $\mathbb{P}_x(\tau_y < \infty) = 1$. The two statements about $\mathbb{E}_x(N_y)$ follow from the identity

$$\mathbb{E}_x(N_y) = \mathbb{P}_x(\tau_y < \infty)(1 + \mathbb{E}_y(N_y)),$$

which itself follows from the strong Markov property, and is valid for both recurrent and transient chains.   □

Remark 2.2 Transience does not imply that $\mathbb{P}_x(\tau_y < \infty) < 1$ for all x, y. Consider the chain on $\mathbb{N}$ whose transition matrix is given by

$$P(x, x+1) = p \in (\tfrac{1}{2}, 1), \ P(x+1, x) = 1 - p \text{ for all } x \in \mathbb{N} \text{ and } P(0, 0) = 1 - p.$$

By the strong law of large numbers, $\mathbb{P}_x(\tau_y < \infty) = 1$ for all $x < y$ and the chain is transient.

Example 2.3 (Pólya Walks)   The Pólya walk on $\mathbb{Z}^d$ is the Markov chain with transition matrix

$$P(x, y) = \frac{1}{2d}\mathbf{1}_{\{x \sim y\}},$$

where $x \sim y \Leftrightarrow \sum_{i=1}^d |x_i - y_i| = 1$. In 1921, Pólya proved that the associated chain is recurrent for $d \leq 2$ and transient for $d \geq 3$.

The proof for $d = 1$ goes as follows. Clearly

$$P^{2k+1}(0, 0) = 0 \text{ and } P^{2k}(0, 0) = \frac{1}{2^{2k}}\binom{2k}{k}.$$

Stirling's formula ($\ln(n!) = n(\ln(n) - 1) + \tfrac{1}{2}(\ln(n) + \ln(2\pi)) + O(\tfrac{1}{n})$) then yields

$$P^{2k}(0, 0) \sim \frac{1}{\sqrt{2\pi k}}.$$

This proves that $\sum_k P^k(0, 0) = \infty$, hence the recurrence.

For $d = 2$, recurrence can be deduced from Exercise 2.4 below. The proof of transience for $d \geq 3$ is slightly more involved and can be found in classical textbooks (see, e.g., [7] or Woess's book [70] for a more advanced textbook on Markov chains on graphs and groups).

Exercise 2.4 (Pólya Walks) Let $X_n = (X_n^1, \ldots, X_n^d)$, where the (X_n^i), $i = 1, \ldots, d$ are independent Pólya walks on $\mathbb{Z}$. Show that (X_n) is recurrent if and only if $d \leq 2$. Deduce from this result the recurrence of the Pólya walk on $\mathbb{Z}^2$.

Exercise 2.5 (Generating Functions) Let $0 < p < 1$ and $q = 1 - p$. Consider the biased walk on $\mathbb{Z}$ whose transition matrix is given by $P(x, x+1) = p$, $P(x, x-1) = q$ and $P(x, y) = 0$ for $|x - y| \neq 1$.
For all $0 \leq t \leq 1$ and $y \in \mathbb{Z}$, set

$$U_y(t) = \mathbb{E}_0(t^{\tau_y} \mathbf{1}_{\{\tau_y < \infty\}})$$

and

$$G_y(t) = \mathbb{E}_0\left(\sum_{k \geq 0} \mathbf{1}_{X_k = y} t^k\right) = \sum_k P^k(0, y) t^k.$$

(i) Prove the following identities:

$$U_0(t) = t(pU_{-1}(t) + qU_1(t)), \ U_1(t) = t(p + qU_{-2}(t)), \ U_{-1}(t) = t(q + pU_{-2}(t)),$$

$$U_2(t) = U_1^2(t), \ U_{-2}(t) = U_{-1}^2(t),$$

and $G_0(t) = \frac{1}{1 - U_0(t)}$.
(ii) Compute $U_0(t)$, $G_0(t)$ and show that

$$\mathbb{E}_x(N_x) = \frac{1}{|1 - 2p|}, \quad \mathbb{E}_x(\tau_x | \tau_x < \infty) = \left(1 - \frac{1}{2\max(p, q)}\right)^{-1}.$$

Comment on these results.

2.1.1 *Positive Recurrence*

A recurrent point x is called *positive recurrent* if $\mathbb{E}_x(\tau_x) < \infty$ and *null recurrent* otherwise.

A measure (respectively a probability measure) π on M is called *invariant* for a transition matrix P if $\pi P = P$, or equivalently,

$$\pi(x) = \sum_{y \in M} \pi(y)P(y, x)$$

for all $x \in M$. Here, we write $\pi(x)$ instead of $\pi(\{x\})$ to highlight the link with matrix-vector notation. Precisely, if $M = \{1, \ldots, N\}$ or $M = \mathbb{N}^*$, and if $x \in M$, then $\pi(x)$ is the xth entry of the row vector $(\pi(\{1\}), \pi(\{2\}), \ldots, \pi(\{N\}))$ or $(\pi(\{1\}), \pi(\{2\}), \ldots)$. If π is an invariant probability measure for P and if X_0 is distributed according to π, then X_n is distributed according to π for all $n \geq 1$ by Proposition 1.6.

The next result shows that for an irreducible recurrent kernel, either all points are positive recurrent or all points are null recurrent. Moreover, positive recurrence is equivalent to the existence of an invariant probability measure.

Theorem 2.6 *Suppose P is irreducible. Then the following assertions are equivalent:*

(a) *There exists an invariant probability measure π for P;*
(b) *There exists a positive recurrent point.*

Under these equivalent conditions:

 (i) *All the points are positive recurrent;*
(ii) *For every initial probability distribution ν on M and $x \in M$,*

$$\lim_{n \to \infty} \frac{1}{n} \sum_{k=1}^{n} 1_{\{X_k = x\}} = \pi(x) = \frac{1}{\mathbb{E}_x(\tau_x)}$$

   $\mathbb{P}_\nu$-*a.s. (in particular, π is unique);*
(iii) *For all $x \in M$ and $f : M \to \mathbb{R}$ bounded or $f : M \to [0, \infty]$,*

$$\pi f = \frac{\mathbb{E}_x(\sum_{k=0}^{\tau_x - 1} f(X_k))}{\mathbb{E}_x(\tau_x)};$$

(iv) *For all $x, y \in M$, $\mathbb{E}_y(\tau_x) < \infty$.*

Proof For all $x \in M$, $\sum_{k=1}^{n} 1_{\{X_k = x\}} = 1_{\{\tau_x < \infty\}} \sum_{k=\tau_x}^{n} 1_{\{X_k = x\}}$. Then, using irreducibility and Proposition 2.1 (ii), one has for every probability measure ν on M

$$\lim_{n \to \infty} \frac{\sum_{k=1}^{n} 1_{\{X_k = x\}}}{n} = \frac{1_{\{\tau_x < \infty\}}}{\mathbb{E}_x(\tau_x)} \tag{2.1}$$

$\mathbb{P}_\nu$-a.s., with the convention that the right-hand term is zero if x is transient. Suppose now that π is an invariant probability measure. By irreducibility and the relation

$\pi(x) = \sum_y \pi(y)P(y,x)$, one sees that $\pi(x) > 0$ for all $x \in M$. Taking $\mathbb{E}_\pi$-expectation on both sides of (2.1) and using dominated convergence gives

$$0 < \pi(x) = \frac{\mathbb{P}_\pi(\tau_x < \infty)}{\mathbb{E}_x(\tau_x)}.$$

This implies $\mathbb{E}_x(\tau_x) < \infty$ so that x is positive recurrent. By Proposition 2.1 (iii), recurrence implies $\mathbb{P}_\pi(\tau_x < \infty) = 1$. Thus $\pi(x) = \frac{1}{\mathbb{E}_x(\tau_x)}$. Suppose now that there exists a positive recurrent point x. Let π be the probability measure defined as in assertion (iii) of Theorem 2.6. We claim that π is an invariant probability measure (compare with Exercise 4.24). For all $f \in B(M)$,

$$\mathbb{E}_x(\tau_x)\,\pi f = \mathbb{E}_x\left(\sum_{k\geq 0} \mathbf{1}_{\{k<\tau_x\}} f(X_k)\right) = \mathbb{E}_x\left(\sum_{k\geq 0} \mathbf{1}_{\{k<\tau_x\}} f(X_{k+1})\right)$$

because $f(X_{\tau_x}) = f(x)$. Thus, using the Markov property and Fubini's theorem,

$$\mathbb{E}_x(\tau_x)\,\pi f = \sum_{k\geq 0} \mathbb{E}_x(\mathbb{E}(f(X_{k+1})\mathbf{1}_{\{k<\tau_x\}}|\mathcal{F}_k))$$

$$= \mathbb{E}_x\left(\sum_{k\geq 0} \mathbf{1}_{\{k<\tau_x\}} Pf(X_k)\right) = \mathbb{E}_x(\tau_x)\pi(Pf).$$

This shows that $\pi Pf = \pi f$, hence $\pi P = \pi$.

It remains to prove assertion (iv). Let $x \neq y \in M$. By irreducibility one can choose $k \geq 1$ such that $P^k(x,y) > 0$. Let $\tau_{k,x} := \inf\{n \geq k : X_n = x\}$. Then $\tau_{k,x} \leq \tau_x^{(k)}$ and, consequently,

$$k + \mathbb{E}_x(\mathbb{E}_{X_k}(\tau_x)\mathbf{1}_{\{X_k\neq x\}}) = \mathbb{E}_x(\tau_{k,x}) \leq \mathbb{E}_x(\tau_x^{(k)}) = \frac{k}{\pi(x)}.$$

Here the last equality follows from assertion (ii) and the strong Markov property. By the Markov property,

$$\mathbb{E}_x(\tau_{k,x}) = k + \mathbb{E}_x(\mathbb{E}_{X_k}(\tau_x \mathbf{1}_{\{X_k\neq x\}})) \geq k + P^k(x,y)\mathbb{E}_y(\tau_x).$$

This shows that

$$\mathbb{E}_y(\tau_x) \leq \frac{k(1-\pi(x))}{\pi(x)P^k(x,y)} < \infty.$$

$\square$

An irreducible kernel (or chain) satisfying one of the equivalent conditions (a) or (b) of Theorem 2.6 is called a positive recurrent kernel (chain).

Corollary 2.7 *If M is finite and P is irreducible, then P is positive recurrent.*

Proof The set $\mathcal{P}(M)$ of probability measures on M is nothing but the unit simplex in $\mathbb{R}^d$ with d the cardinality of M. By Brouwer's fixed point theorem (see, e.g., Corollary XVI.2.2 in [23]), the map $\mathcal{P}(M) \ni \pi \mapsto \pi P \in \mathcal{P}(M)$ has a fixed point, which is then an invariant probability measure for P.   □

Remark 2.8 The proof of Corollary 2.7 shows that every Markov chain on a finite set, possibly non-irreducible, always admits (at least) one invariant probability measure.

Exercise 2.9 Give a direct proof of this latter fact. *Hint:* Consider the sequence (μ_n) defined by $\mu_n = \frac{1}{n} \sum_{k=1}^{n} \mu P^k$, where μ is some probability measure.

An interesting consequence of Theorem 2.6 *(iii)* is the next proposition, which relates moments of the first return time to x to π-mean moments of the hitting time of x.

Proposition 2.10 *Suppose P is positive recurrent with invariant probability measure π. Then, for every nonnegative function $\psi : \mathbb{N} \to \mathbb{R}_+$ and every $x \in M$,*

$$\mathbb{E}_\pi(\psi(\tau_x)) = \pi(x)\mathbb{E}_x\left(\sum_{k=1}^{\tau_x} \psi(k)\right).$$

In particular, for every $\lambda > 0$,

$$\mathbb{E}_\pi(e^{\lambda \tau_x}) = \pi(x)\frac{e^\lambda}{e^\lambda - 1}[\mathbb{E}_x(e^{\lambda \tau_x}) - 1];$$

And for every $p \geq 0$,

$$\mathbb{E}_\pi(\tau_x^p) \leq \pi(x)\frac{\mathbb{E}_x[(\tau_x + 1)^{p+1}] - 1}{p + 1}.$$

Proof Fix $\psi : \mathbb{N} \to \mathbb{R}_+$ and $x \in M$. By Theorem 2.6 *(iii)* applied to $f(y) := \mathbb{E}_y(\psi(\tau_x))$, one has

$$\mathbb{E}_\pi(\psi(\tau_x)) = \pi(x)\mathbb{E}_x\left(\sum_{k\geq 0} \mathbf{1}_{\tau_x > k}\mathbb{E}_{X_k}(\psi(\tau_x))\right) = \pi(x)\sum_{k\geq 0}\mathbb{E}_x(\mathbf{1}_{\tau_x > k}\mathbb{E}_{X_k}(\psi(\tau_x))).$$

But, by the Markov property,

$$\mathbb{E}_x(\mathbf{1}_{\tau_x > k}\mathbb{E}_{X_k}(\psi(\tau_x))) = \mathbb{E}_x(\mathbb{E}_x(\psi(\tau_x - k)\mathbf{1}_{\tau_x > k}|\mathcal{F}_k)) = \mathbb{E}_x(\psi(\tau_x - k)\mathbf{1}_{\tau_x > k}).$$

This proves the result.   □

Exercise 2.11 (Pólya Walks, Continued) Show that the Pólya walks on $\mathbb{Z}$ and $\mathbb{Z}^2$ are null recurrent. *Hint:* Show that they do not have any invariant probability measure.

Exercise 2.12 (Reflected Walks) Let $0 < p < 1, q = 1 - p$ and $0 \le r < 1$. Consider the chain on $\mathbb{N}$ whose transition matrix is given by $P(x, x + 1) = p$, $P(x, x - 1) = q$ if $x \ge 1$, $P(0, 0) = r$ and $P(0, 1) = 1 - r$. With the notation of Exercise 2.5 compute $U_0(t)$ and show that the chain is transient for $p > 1/2$, null recurrent for $p = 1/2$ and positive recurrent for $p < 1/2$. Compute $\mathbb{E}_0(\tau_0 | \tau_0 < \infty)$.

Exercise 2.13 (Harmonic Functions) A function $h : M \to \mathbb{R}$ is called harmonic for the Markov kernel P if $Ph = h$. Suppose P is irreducible and recurrent. Show that every nonnegative or bounded harmonic function is constant. (*Hint:* Show that $h(X_n)$ is a nonnegative (or bounded) martingale, hence convergent by Theorem A.6.) Give an example of a nonconstant unbounded harmonic function for the Pólya walk on $\mathbb{Z}$.

Exercise 2.14 (Reversibility) Let π be a probability measure on M. A Markov kernel P is said to be *reversible* with respect to π if $\pi(x)P(x, y) = \pi(y)P(y, x)$ for all $x, y \in M$.

 (i) Show that if P is reversible with respect to π, then π is invariant for P.
 (ii) Show that if P is reversible with respect to π and if $\pi(x) > 0$ for all $x \in M$, then $Pf(x) := \sum_{y \in M} P(x, y)f(y)$ defines a self-adjoint operator on the Hilbert space $l^2(\pi) := \{f : M \to \mathbb{R} : \sum_{x \in M} \pi(x)|f(x)|^2 < \infty\}$ with inner product $\langle f, g \rangle := \sum_{x \in M} \pi(x)f(x)g(x)$, i.e., $\langle Pf, g \rangle = \langle f, Pg \rangle$ for all $f, g \in l^2(\pi)$.
 (iii) Give an example of a Markov kernel P and a probability measure π such that π is invariant for P, but P is not reversible with respect to π.

2.1.2 Null Recurrence

Although an irreducible null recurrent chain has no invariant probability measure (for otherwise it would be positive recurrent) it always has an unbounded invariant measure.

Theorem 2.15 *Suppose P is irreducible and null recurrent. Given $x \in M$, let π be the measure on M defined by*

$$\pi f = \mathbb{E}_x \left(\sum_{k=0}^{\tau_x - 1} f(X_k) \right)$$

for $f : M \to \mathbb{R}$ *nonnegative. Then* π *is* σ*-finite* ($\pi(y) < \infty$ *for all* $y \in M$), *positive* ($\pi(y) > 0$ *for all* $y \in M$), *unbounded* ($\pi(M) = \infty$), *and invariant under* P ($\pi = \pi P$). *Every other* σ*-finite invariant measure is proportional to* π.

Proof For $y \neq x$, set $N_{y<x} = \sum_{k=0}^{\tau_x - 1} 1_{\{X_k = y\}}$. By the strong Markov property, for all $k \geq 0$,

$$\mathbb{P}_x(N_{y<x} \geq k+1) = \mathbb{P}_x(\tau_y^{(k+1)} < \tau_x) = \mathbb{P}_x(\tau_y^{(k)} < \tau_x; \tau_y^{(k+1)} < \tau_x)$$

$$= \mathbb{P}_x(\tau_y^{(k)} < \tau_x)\mathbb{P}_y(\tau_y < \tau_x) = a^{k+1},$$

where $a = \mathbb{P}_y(\tau_y < \tau_x) < 1$ (by irreducibility). This proves that

$$0 < \pi(y) = \frac{a}{1-a} < \infty.$$

Invariance of π is proved exactly as in Theorem 2.6 (*iii*). Clearly $\pi(M) = \infty$ for otherwise $\frac{\pi}{\pi(M)}$ would be an invariant probability measure, in contradiction with the assumption that the chain is null recurrent.

It remains to show that every other σ-finite invariant measure is proportional to μ. Let $Q(x, y) = \frac{\mu(y)P(y,x)}{\mu(x)}$. Then Q is a Markov kernel and $Q^n(x, y) = \frac{\mu(y)P^n(y,x)}{\mu(x)}$. It follows that Q is also irreducible and null recurrent by application of Proposition 2.1. Let now ν be another σ-finite invariant measure. Then $h(x) = \frac{\nu(x)}{\mu(x)}$ is harmonic for Q, hence constant (see Exercice 2.13). This concludes the proof. $\square$

2.2   Subsets of Recurrent Sets

Given $C \subset M$, we let

$$\tau_C = \tau_C^{(1)} := \inf\{n \geq 1 : X_n \in C\},$$

and

$$\tau_C^{(k+1)} := \inf\{n > \tau_C^{(k)} : X_n \in C\}$$

for all $k \geq 1$. We also set $\tau_C^{(0)} := 0$. The next proposition shows that, whenever P is irreducible, recurrence (respectively positive recurrence) of the chain is equivalent to recurrence (positive recurrence) of any finite subset.

Proposition 2.16 *Suppose P is irreducible and let $C \subset M$ be a nonempty finite set such that for all $x \in C$, $\mathbb{P}_x(\tau_C < \infty) = 1$ (respectively $\mathbb{E}_x(\tau_C) < \infty$). Then P is recurrent (respectively positive recurrent).*

Proof Let $x \in C$. Then, since $\mathbb{P}_y(\tau_C < \infty) = 1$ for all $y \in C$, the strong Markov property implies that (X_n) visits C infinitely often $\mathbb{P}_x$-almost surely. Since C is finite, it follows that $\mathbb{P}_x$-almost surely, there is $y \in C$ such that $N_y = \infty$. If P was transient, we would have by Proposition 2.1 that $\mathbb{P}_x(\bigcup_{y \in C}\{N_y = \infty\}) = 0$, a contradiction. Hence P is recurrent.

Suppose now that $K := \max_{x \in C} \mathbb{E}_x(\tau_C) < \infty$. Let Q be the Markov kernel on C defined by $Q(x, y) := \mathbb{P}_x(X_{\tau_C} = y)$ for $x, y \in C$. Since C is finite, Q admits an invariant probability measure π (see Remark 2.8). Thus, if X_0 has law π, then X_{τ_C} has also law π. It follows (by a proof similar to the proof of Theorem 2.6 (iii) or by Exercise 4.24) that the measure μ defined by

$$\mu f = \frac{\mathbb{E}_\pi(\sum_{k=0}^{\tau_C-1} f(X_k))}{\mathbb{E}_\pi(\tau_C)}$$

is invariant for P. Note here that $\mathbb{E}_\pi(\tau_C) \leq K < \infty$. This proves positive recurrence.   □

Exercise 2.17 Suppose P is irreducible, $C \subset M$ is finite and for all $x \in M \setminus C$, $\mathbb{P}_x(\tau_C < \infty) = 1$. Show that P is recurrent. *Hint:* If $M \setminus C \neq \emptyset$, prove that for all $x \in C$, $\mathbb{P}_x(\tau_{M \setminus C} < \infty) = 1$ and then use Proposition 2.16.

The next result extends and generalizes Proposition 2.16. The second part contains a classical result originally due to Chung [16]. The proof given here is different.

Proposition 2.18 *Suppose P is irreducible and let $C \subset M$ be a finite set.*

(i) *Assume that for some $\lambda_0 > 0$ and all $x \in C$, $\mathbb{E}_x(e^{\lambda_0 \tau_C}) < \infty$. Then, for all $x, y \in M$, there exists $\lambda \in (0, \lambda_0]$ such that*

$$\mathbb{E}_x(e^{\lambda \tau_y}) < \infty.$$

(ii) *Let $p \geq 1$ and suppose that for all $x \in C$, $\mathbb{E}_x(\tau_C^p) < \infty$. Then, for all $x, y \in M$,*

$$\mathbb{E}_x(\tau_y^p) < \infty.$$

Proof

(i) First assume that $M = C$. In this case there exists, by irreducibility, some $\varepsilon > 0$ such that for all $x, y \in M$ and $k := \mathsf{card}(M)$, $\mathbb{P}_x(\tau_y > k) \leq 1 - \varepsilon$. Therefore, by the Markov property and induction on $n \geq 1$,

$$\mathbb{P}_x(\tau_y > nk) = \mathbb{E}_x(1_{\tau_y > (n-1)k})\mathbb{P}_{X_{(n-1)k}}(\tau_y > k) \leq (1 - \varepsilon)^n.$$

Thus, for all $n \geq 0$,

$$\mathbb{P}_x(\tau_y > n) \leq \mathbb{P}_x(\tau_y > k[\tfrac{n}{k}]) \leq (1 - \varepsilon)^{\frac{n}{k} - 1},$$

where $[\frac{n}{k}]$ is the largest integer less than or equal to $\frac{n}{k}$. Hence, for $\alpha > 0$ so small that $e^{k\alpha}(1 - \varepsilon) < 1$,

$$\mathbb{E}_x(e^{\alpha \tau_y}) \leq \sum_{n=1}^{\infty} e^{\alpha n} \mathbb{P}_x(\tau_y \geq n) < \infty.$$

We now turn to the proof of the first statement in full generality. Let

$$Y_n = X_{\tau_C^{(n)}}.$$

Such a definition makes sense because, by recurrence, $\tau_C^{(n)} < \infty$ almost surely. For all $y \in C$, set $\sigma_y := \inf\{n \geq 1 : Y_n = y\}$. For $x \in C, (Y_n)$ is a C-valued Markov chain on the probability space $(M^{\mathbb{N}}, \mathcal{B}(M^{\mathbb{N}}), \mathbb{P}_x)$, with respect to the filtration $\{\mathcal{F}_{\tau_C^{(n)}}\}_n$, and with Markov kernel $Q(a, b) := \mathbb{P}_a(X_{\tau_C} = b)$ introduced in the proof of Proposition 2.16. Thus, by what precedes,

$$\max_{x,y \in C} \mathbb{E}_x(e^{\alpha \sigma_y}) < \infty \qquad (2.2)$$

for some $\alpha > 0$.

By assumption, $\max_{x \in C} \mathbb{E}_x(e^{\lambda_0 \tau_C}) \leq e^{\alpha_0}$ for some $\alpha_0 \geq 0$. By Jensen's inequality, for all $t \in [0, 1], \mathbb{E}_x(e^{t\lambda_0 \tau_C}) \leq \mathbb{E}_x(e^{\lambda_0 \tau_C})^t \leq e^{t\alpha_0}$. Choose $\lambda \in (0, \frac{\lambda_0}{2}]$ so small that $2\lambda\alpha_0 \leq \lambda_0 \alpha$. Then

$$\max_{x \in C} \mathbb{E}_x(e^{2\lambda \tau_C}) \leq e^{\alpha}.$$

Set $M_n := e^{(2\lambda \tau_C^{(n)} - n\alpha)}$. The previous inequality combined with the strong Markov property shows that (M_n) is a supermartingale under $\mathbb{P}_x$ with respect to the filtration $\{\mathcal{F}_{\tau_C^{(n)}}\}_n$. Therefore, using Theorem A.4 on optional stopping, $(M_{n \wedge \sigma_y})$ is again a supermartingale, and in particular $\mathbb{E}_x(M_{n \wedge \sigma_y}) \leq \mathbb{E}_x(M_0) = 1$. Together with Hölder's inequality, this yields for all $x, y \in C$

$$\mathbb{E}_x(e^{\lambda \tau_C^{(n \wedge \sigma_y)}}) \leq \mathbb{E}_x(M_{n \wedge \sigma_y})^{1/2} \mathbb{E}_x(e^{\alpha(n \wedge \sigma_y)})^{1/2} \leq \mathbb{E}_x(e^{\alpha \sigma_y})^{1/2} < \infty.$$

Thus,

$$\mathbb{E}_x(e^{\lambda \tau_y}) = \mathbb{E}_x(e^{\lambda \tau_C^{(\sigma_y)}}) < \infty$$

for all $x, y \in C$.

In order to conclude the proof, it suffices to show that for any finite set C' containing C, $\max_{x \in C'} \mathbb{E}_x(e^{\lambda_0 \tau_C}) < \infty$. Then, by what precedes (with C' in place of C), this will imply that $\max_{x,y \in C'} \mathbb{E}_x(e^{\lambda' \tau_y}) < \infty$ for some $\lambda' \in (0, \lambda_0]$.

We reason like in the proof of Theorem 2.6 (iv). Let $C' \supseteq C$, $y \in C' \backslash C$. Fix $x \in C$. Then, for some $k \geq 1$, $P^k(x, y) > 0$. Let $\tau_{k,C} = \min\{n \geq k : X_n \in C\}$. One has $\tau_{k,C} \leq \tau_C^{(k)}$. Thus,

$$e^{\lambda_0 k} P^k(x, y) \mathbb{E}_y(e^{\lambda_0 \tau_C}) \leq \mathbb{E}_x(e^{\lambda_0 k} \mathbb{E}_{X_k}(e^{\lambda_0(\tau_C)} \mathbf{1}_{X_k \notin C})) = \mathbb{E}_x(e^{\lambda_0 \tau_{k,C}} \mathbf{1}_{X_k \notin C})$$

$$\leq \mathbb{E}_x(e^{\lambda_0 \tau_C^{(k)}}) \leq [\max_{z \in C} \mathbb{E}_z(e^{\lambda_0 \tau_C})]^k < \infty.$$

This concludes the proof of (i).

(ii) Slightly adapting the previous argument, one easily shows that

$$\max_{x \in C} \mathbb{E}_x(\tau_C^p) < \infty \implies \max_{x \in C'} \mathbb{E}_x(\tau_C^p) < \infty$$

for any finite set C' containing C. It then suffices to show that, for all $x, y \in C$, $\mathbb{E}_x(\tau_y^p) < \infty$.

By the assumption and the strong Markov property, there exists $K \geq 0$ such that for every $n \geq 0$,

$$\mathbb{E}_x(|\tau_C^{(n+1)} - \tau_C^{(n)}|^p | \mathcal{F}_{\tau_C^{(n)}}) = \mathbb{E}_{Y_n}(\tau_C^p) \leq K^p.$$

Therefore, with $\| \cdot \|_p = \mathbb{E}_x(| \cdot |^p)^{1/p}$,

$$\|\tau_y\|_p = \|\tau_C^{(\sigma_y)}\|_p = \left\| \sum_{i \geq 0} (\tau_C^{(i+1)} - \tau_C^{(i)}) \mathbf{1}_{i < \sigma_y} \right\|_p \leq \sum_{i \geq 0} \|(\tau_C^{(i+1)} - \tau_C^{(i)}) \mathbf{1}_{i < \sigma_y}\|_p.$$

Now

$$\mathbb{E}_x(|\tau_C^{(i+1)} - \tau_C^{(i)}|^p \mathbf{1}_{i < \sigma_y}) = \mathbb{E}_x(\mathbb{E}_x(|\tau_C^{(i+1)} - \tau_C^{(i)}|^p | \mathcal{F}_{\tau_C^{(i)}}) \mathbf{1}_{i < \sigma_y}) \leq K^p \mathbb{P}_x(\sigma_y > i).$$

Thus

$$\|\tau_y\|_p \leq K \sum_{i \geq 0} \mathbb{P}_x(\sigma_y > i)^{1/p} < \infty,$$

because, as seen in the beginning of the proof, the law of σ_y has a geometric tail.

$\square$

2.3 Recurrence and Lyapunov Functions

By Proposition 2.1, the divergence (respectively convergence) of the series $\sum_{k \geq 1} P^k(x, x)$ is a criterion for the recurrence (transience) of the point x, but such a criterion may be difficult to verify in practice. We discuss here other criteria based on Lyapounov functions, a tool that will play a key role in the next chapters. In brief, a *Lyapunov function* is a map $V : M \to [1, \infty)$ such that $PV - V \leq 0$ outside a certain subset $C \subset M$. Lyapunov functions are practical tools to ensure that the assumptions of Propositions 2.16 and 2.18 are satisfied.

A map $V : M \to \mathbb{R}_+$ is called *proper* if for every $R > 0$, the set $\{x \in M : V(x) \leq R\}$ is finite. If M is finite, every map $V : M \to \mathbb{R}_+$ is proper. If M is countably infinite and $(x_n)_{n \geq 1}$ is any enumeration of the elements of M, $V : M \to \mathbb{R}_+$ is proper if and only if $\lim_{n \to \infty} V(x_n) = \infty$.

Apart from the first assertion, the following result is a consequence of a more general result (Proposition 7.12) that will be proved later.

Theorem 2.19 *Let P be a Markov kernel, let $V : M \to [1, \infty)$ be a map, and let $C \subset M$ be nonempty. Consider the following conditions:*

(a) P is irreducible, $PV - V \leq 0$ on $M \setminus C$ and V is proper;
(b) $PV - V \leq -1$ on $M \setminus C$ and $PV < \infty$ on C;
(b') Condition (b) and in addition

$$\sup_{x \in M} \mathbb{E}_x(|V(X_1) - V(x)|^p) < \infty$$

for some $p \geq 1$;
(c) $PV - V \leq -\lambda V$ on $M \setminus C$ for some $\lambda \in (0, 1)$ and $PV < \infty$ on C.

Then, for all $x \in M$,

  (i) Under condition (a),

$$\mathbb{P}_x(\tau_C < \infty) = 1;$$

  (ii) Under condition (b),

$$\mathbb{E}_x(\tau_C) \leq PV(x) + 1;$$

  (iii) Under condition (b'),

$$\mathbb{E}_x(\tau_C^p) \leq c(1 + V(x)^p)$$

  for some constant $c > 0$ that depends on p but does not depend on x;
  (iv) Under condition (c),

$$\mathbb{E}_x(e^{\lambda \tau_C}) \leq \mathbb{E}_x(e^{-\log(1-\lambda)\tau_C}) \leq \frac{1}{1-\lambda} PV(x).$$

In particular, if P is irreducible and if C is finite, conditions $(a), (b), (b'), (c)$ respectively ensure recurrence of P, positive recurrence of P, p-th moments for the hitting times τ_y under $\mathbb{P}_x$, and exponential moments for τ_y under $\mathbb{P}_x$ for every $x, y \in M$.

Proof We only prove the first assertion. The other three follow from Proposition 7.12 to be proved later. When P is irreducible and when C is finite, recurrence, positive recurrence, p-th moments, and exponential moments of hitting times are direct consequences of Propositions 2.16 and 2.18.

By irreducibility, the chain is either recurrent or transient. If it is recurrent, $\mathbb{P}_x(\tau_C < \infty) = 1$ for every $x \in M$ by Proposition 2.1. Suppose the chain is transient. For $x \in M \setminus C$, the sequence $V_n := V(X_{n \wedge \tau_C})$ is under $\mathbb{P}_x$ a supermartingale because $\mathbb{E}_x(V_{n+1} - V_n | \mathcal{F}_n) = (PV(X_n) - V(X_n))\mathbf{1}_{\tau_C > n} \leq 0$. Thus, being nonnegative, (V_n) converges $\mathbb{P}_x$-almost surely to some random variable V_∞ taking values in $[0, \infty)$ (apply Theorem A.6 to the submartingale $(-V_n)$). This shows that $V(X_n)$ converges $\mathbb{P}_x$-almost surely on $\{\tau_C = \infty\}$. On the other hand, by transience (Proposition 2.1 (iii)) and by the assumption that V is proper, $\limsup_{n \to \infty} V(X_n) = \infty$ $\mathbb{P}_x$-almost surely, and therefore $\mathbb{P}_x(\tau_C < \infty) = 1$. And for $x \in C$, we have by the Markov property

$$\mathbb{P}_x(\tau_C < \infty) = \mathbb{P}_x(X_1 \in C) + \mathbb{E}_x(\mathbf{1}_{X_1 \in M \setminus C}\mathbb{P}_{X_1}(\tau_C < \infty)) = 1.$$

$\square$

Exercise 2.20 Suppose $V : M \to [1, \infty)$ is a proper map. Show that condition (c) in Theorem 2.19 for a nonempty finite set C is equivalent to the existence of constants $0 \le \rho < 1$ and $\kappa \ge 0$ such that

$$PV \le \rho V + \kappa.$$

Show that under such a condition, every invariant probability measure π satisfies

$$\pi V \le \frac{\kappa}{1 - \rho} < \infty.$$

See Corollary 4.23 for a proof of the second assertion.

2.4   Aperiodic Chains

We start with a general definition of aperiodicity. Let $R \subset \mathbb{N}^*$ be a (nonempty) set closed under addition. That is

$$i, j \in R \Rightarrow i + j \in R.$$

The *period* of R is defined as its greatest common divisor. If this period is 1, R is said to be *aperiodic*. Aperiodic sets enjoy the following useful property, that will be used repeatedly throughout the book.

Proposition 2.21 *Let R be aperiodic. Then there exists $n_0 \in \mathbb{N}$ such that $n_0 + \mathbb{N} = \{n \in \mathbb{N} : n \ge n_0\} \subset R$.*

Proof There exist, by aperiodicity, $a_1, \ldots, a_l \in R$ whose greatest common divisor is 1. (To see this, take any element of R and call it a_1; then a_1 has a finite number of divisors strictly greater than 1, which we denote by $d_2, \ldots, d_l$; for $2 \le i \le l$, pick a_i from R such that d_i does not divide a_i; such a_i exists because the greatest common divisor of R is 1). By Bézout's identity, there exist $q_1, \ldots q_l \in \mathbb{Z}$ such that $\sum_i q_i a_i = -1$. Set $a := \sum_{i:q_i>0} q_i a_i$. The set R being closed under addition, both a and $a + 1 = \sum_{i:q_i<0} -q_i a_i$ lie in R. Every $n \ge a^2$ can be written as $n = ka + r = (k - r)a + r(a + 1)$ for some $r \in \{0, \ldots, a - 1\}$ and $k \ge a$. Thus, every $n \ge a^2$ is an element of R. $\square$

We now turn to the definition of aperiodicity for a countable Markov chain. Given a kernel P on M and $x \in M$, let $R(x) := \{k \ge 1 : x \leadsto^k x\}$ be the set of possible return times to x. The *period* of x, $per(x)$, is defined as the period of $R(x)$ and x is called *aperiodic* whenever $R(x)$ is. The kernel (or the chain) is said to be aperiodic if all points $x \in M$ are aperiodic.

Proposition 2.22 *Suppose P is irreducible. Then*

(i) *All points $x \in M$ have the same period;*
(ii) *P is aperiodic if and only if for all $x, y \in M$ there exists $n(x, y) \in \mathbb{N}$ such that $x \rightsquigarrow^n y$ for all $n \geq n(x, y)$.*

Proof

(i) Let $x, y \in M$. By irreducibility, there exist $i, j \in \mathbb{N}^*$ such that $x \rightsquigarrow^i y$ and $y \rightsquigarrow^j x$. Thus $i + j \in R(x)$ and for all $k \in R(y), i + j + k \in R(x)$. Therefore, $per(x)$ divides $i + j$ and $i + j + k$, hence k, for all $k \in R(y)$. Thus $per(x) \leq per(y)$ and by symmetry $per(x) = per(y)$.

(ii) The "if" part is obvious. We prove the "only if" part. Given $y \in M$, there exists, by Proposition 2.21, $n_0 \in \mathbb{N}$ such that $n \in R(y)$ for all $n \geq n_0$. If now x is another point in M, $x \rightsquigarrow^i y$ for some i by irreducibility, hence $x \rightsquigarrow^n y$ for all $n \geq n_0 + i$.

$\square$

An immediate useful consequence of Proposition 2.22 is the next result. Given two Markov kernels P and $\tilde{P}$ respectively defined on the countable state space M and $\tilde{M}$, we let $P \otimes \tilde{P}$ denote the Markov kernel on $M \times \tilde{M}$ corresponding to two independent chains with kernels $P, \tilde{P}$. That is

$$(P \otimes \tilde{P})((x, x'); (y, y')) := P(x, y)\tilde{P}(x', y').$$

Corollary 2.23 *If P and $\tilde{P}$ are both irreducible and aperiodic, so is $P \otimes \tilde{P}$. If in addition P and $\tilde{P}$ are positive recurrent, so is $P \otimes \tilde{P}$.*

Proof Note that $(P \otimes \tilde{P})^n = P^n \otimes \tilde{P}^n$ for every $n \in \mathbb{N}^*$. Thus, irreducibility (and aperiodicity) of $P \otimes \tilde{P}$ follows from Proposition 2.22 (ii), applied to P and $\tilde{P}$. Also, if π and $\tilde{\pi}$ are invariant probability measures for P and $\tilde{P}$, so is $\pi \otimes \tilde{\pi}$ (defined as $((\pi \otimes \tilde{\pi})(x, x') := \pi(x)\tilde{\pi}(x'))$ for $P \otimes \tilde{P}$. By Theorem 2.6, this proves positive recurrence. $\square$

Exercise 2.24 Give an example of an irreducible and positive recurrent kernel P such that $P \otimes P$ is not irreducible, and an example of an irreducible recurrent kernel P such that $P \otimes P$ is irreducible and transient.

Exercise 2.25 Show that if $P \otimes \tilde{P}$ is irreducible, then both P and $\tilde{P}$ are irreducible. Also show that if $P \otimes \tilde{P}$ is irreducible and recurrent, then both P and $\tilde{P}$ are recurrent.

Exercise 2.26 Let $(X_n)_{n \geq 0}$ be a Markov chain on $\mathbb{Z} \setminus \{0\}$ whose transition matrix P is given by

$$P(i, i + 1) = P(i, -i) = 1/2, \ i \geqslant 1$$
$$P(-1, 1) = P(i, i + 1) = 1, \ i \leqslant -2.$$

 (i) Draw the weighted directed graph associated with (X_n) and determine whether
      the chain is irreducible.
 (ii) Find the period of the chain.
(iii) Find a Lyapunov function V and a finite set $C \subset \mathbb{Z} \setminus \{0\}$ such that P, V, and
      C satisfy condition (b) of Theorem 2.19.
(iv) Show that $(X_n)_{n \geq 0}$ is positive recurrent and find its unique invariant probability
      measure.

2.5   The Convergence Theorem

The main result of this section is the convergence theorem for irreducible aperiodic
Markov chains. This theorem is sometimes called the ergodic theorem in the
literature, but we prefer to reserve this terminology for Birkhoff's ergodic theorem.

Theorem 2.27 *Suppose P is irreducible and aperiodic. Let μ be a probability
measure on M.*

 (i) *If P is positive recurrent with invariant probability measure π, then*

$$\lim_{n \to \infty} \sup_{z \in M} |\mu P^n(z) - \pi(z)| = 0.$$

(ii) *If P is not positive recurrent, then, for all $z \in M$,*

$$\lim_{n \to \infty} \mu P^n(z) = 0.$$

Proof Let $(X_n, Y_n)_{n \in \mathbb{N}}$ be the canonical chain on $(M \times M)^{\mathbb{N}}$ (i.e., $(X_n, Y_n)(\omega, \tilde{\omega}) :=
(\omega_n, \tilde{\omega}_n)$), and let

$$\tau_\Delta := \inf\{n \geq 1 : (X_n, Y_n) \in \Delta\},$$

where $\Delta := \{(x, x) : x \in M\}$ is the diagonal of M. Throughout the proof, we write
$\mathbb{P}_\alpha$ (respectively $\mathbb{P}_{x,y}$) for the Markov measure on $(M \times M)^{\mathbb{N}}$ with kernel $P \otimes P$
and initial distribution α (respectively $\delta_{x,y}$). By Corollary 2.23, $P \otimes P$ is irreducible,
hence either recurrent or transient.

Case 1 $P \otimes P$ is recurrent. For all $x, y, z \in M$,

$$\mathbb{P}_{x,y}(X_n = z) = \mathbb{P}_{x,y}(X_n = z; \tau_\Delta > n) + \mathbb{P}_{x,y}(X_n = z; \tau_\Delta \leq n)$$

$$= \mathbb{P}_{x,y}(X_n = z; \tau_\Delta > n) + \mathbb{P}_{x,y}(Y_n = z; \tau_\Delta \leq n)$$

$$\leq \mathbb{P}_{x,y}(\tau_\Delta > n) + \mathbb{P}_{x,y}(Y_n = z),$$

where the second equality follows from the strong Markov property and the fact that $X_{\tau_\Delta} = Y_{\tau_\Delta}$. Interchanging the roles of X_n and Y_n, one also has

$$\mathbb{P}_{x,y}(Y_n = z) \leq \mathbb{P}_{x,y}(\tau_\Delta > n) + \mathbb{P}_{x,y}(X_n = z).$$

Hence

$$|P^n(x, z) - P^n(y, z)| = |\mathbb{P}_{x,y}(X_n = z) - \mathbb{P}_{x,y}(Y_n = z)| \leq \mathbb{P}_{x,y}(\tau_\Delta > n),$$

and by integration

$$|\mu P^n(z) - \nu P^n(z)| \leq \mathbb{P}_{\mu \otimes \nu}(\tau_\Delta > n) \tag{2.3}$$

for every probability measure ν on M and every $z \in M$. By recurrence of $P \otimes P$ (and Proposition 2.1 (iii)), one has for every $x, y \in M$ that $\mathbb{P}_{x,y}(\tau_\Delta > n) \to 0$ as $n \to \infty$. Thus

$$\lim_{n \to \infty} \sup_{z \in M} |\mu P^n(z) - \nu P^n(z)| = 0 \tag{2.4}$$

by dominated convergence. In light of Exercise 2.25, there are two subcases: P is either positive recurrent or null recurrent. If P is positive recurrent, (2.4) applied to $\nu = \pi$, the invariant probability measure of P, proves part (i) of the theorem. If P is null recurrent, let π be an unbounded invariant measure of P (see Theorem 2.15). For any nonempty finite set $A \subset M$, set $\pi_A(x) := \frac{\pi(x) 1_A(x)}{\pi(A)}$. Then $\pi_A \leq \frac{\pi}{\pi(A)}$, whence

$$\pi_A P^n(z) \leq \frac{\pi P^n(z)}{\pi(A)} = \frac{\pi(z)}{\pi(A)}.$$

Therefore, by (2.4) applied to $\nu = \pi_A$,

$$\limsup_{n \to \infty} \mu P^n(z) \leq \lim_{n \to \infty} |\mu P^n(z) - \pi_A P^n(z)| + \frac{\pi(z)}{\pi(A)} = \frac{\pi(z)}{\pi(A)}.$$

Letting $A \uparrow M$ proves (ii) in this case because $\pi(M) = \infty$.

Case 2 $P \otimes P$ is transient. By Proposition 2.1 (i),

$$[P^n(z, z)]^2 = (P \otimes P)^n((z, z); (z, z)) \to 0$$

as $n \to \infty$, for all $z \in M$. By irreducibility of P, this implies that $P^n(x, z) \to 0$ for all $x, z \in M$. Thus $\mu P^n(z) \to 0$ by dominated convergence. This proves (ii) in case 2.

$\square$

As shown below, the convergence in Theorem 2.27 is geometric if there exists a proper map that satisfies condition (c) of Theorem 2.19 for a nonempty finite set C (see also Exercise 2.20).

Theorem 2.28 *Suppose P is irreducible and aperiodic, and that there exists a proper map $V : M \to [1, \infty)$ and constants $0 \le \rho < 1, \kappa \ge 0$ such that*

$$PV \le \rho V + \kappa.$$

Then P is positive recurrent and, denoting by π its invariant probability measure:

(i) *One has $\pi V \le \frac{\kappa}{1-\rho} < \infty$;*
(ii) *There exist constants $0 \le \gamma < 1$ and $c \ge 0$ such that for every probability measure μ on M,*

$$\sup_{z \in M} |\mu P^n(z) - \pi(z)| \le c\gamma^n(\mu V + 1), \quad \forall n \in \mathbb{N}.$$

Corollary 2.29 *Suppose M is finite and P irreducible and aperiodic, with invariant probability measure π. Then there exist constants $0 \le \gamma < 1$ and $c \ge 0$ such that for every probability measure μ on M,*

$$\sup_{z \in M} |\mu P^n(z) - \pi(z)| \le c\gamma^n, \quad \forall n \in \mathbb{N}.$$

Proof Take $V \equiv 1$ in Theorem 2.28. □

Proof *(Of Theorem 2.28)* We use the same notation, $P \otimes P$, (X_n, Y_n), Δ, etc., as in the proof of Theorem 2.27.

Positive recurrence follows from Exercise 2.20 and Theorem 2.19. Assertion (i) follows from Exercise 2.20. By inequality (2.3) from the proof of Theorem 2.27, it suffices to derive an exponential upper bound on $\mathbb{P}_{\mu \otimes \pi}(\tau_\Delta > n)$ in order to prove assertion (ii). Pick $x^* \in M$ and choose $\varepsilon > 0$ small enough so that $V(x^*) \le \frac{\kappa}{\varepsilon}$ and $\rho + \varepsilon < 1$. Set $W(x, y) := V(x) + V(y)$, $x, y \in M$. Then

$$(P \otimes P)W(x, y) = PV(x) + PV(y) \le \rho W(x, y) + 2\kappa,$$

so that $(P \otimes P)W \le (\rho + \varepsilon)W$ on the complement of the set

$$C := \{(x, y) : W(x, y) \le \frac{2\kappa}{\varepsilon}\}.$$

By Theorem 2.19 (iv) and assertion (i), we then obtain, for some positive constant c depending on κ, ρ, and ε,

$$\mathbb{E}_{\mu\otimes\pi}(e^{(1-\rho-\varepsilon)\tau_C}) \leq \frac{(\mu\otimes\pi)(P\otimes P)W}{\rho+\varepsilon} \leq \frac{\rho(\mu V+\pi V)+2\kappa}{\rho+\varepsilon} \leq c(1+\mu V).$$

Since V is proper, the set C is finite, and Proposition 2.18 (i) together with $(x^*, x^*) \in C$ yield the existence of $\lambda > 0$ such that

$$\max_{(x,y)\in C} \mathbb{E}_{(x,y)}(e^{\lambda\tau_{(x^*,x^*)}}) < \infty.$$

Thus

$$\mathbb{P}_{\mu\otimes\pi}(\tau_\Delta > n) \leq \mathbb{P}_{\mu\otimes\pi}(\tau_{(x^*,x^*)} > n)$$

$$\leq \mathbb{P}_{\mu\otimes\pi}(\tau_C > n/2) + \mathbb{E}_{\mu\otimes\pi}(\mathbb{P}_{(X_{\tau_C},Y_{\tau_C})}(\tau_{(x^*,x^*)} > n/2))$$

$$\leq ce^{-\lambda n/2}(1+\mu V)$$

for some other constant c. Inequality (2.3) concludes the proof.   □

2.6 Application to Renewal Theory

Let $(\Delta_i)_{i\geq 1}$ be a sequence of i.i.d. random variables living on some probability space $(\Omega, \mathcal{F}, \mathsf{P})$ and taking values in $\mathbb{N}$. Let Δ_0 be another $\mathbb{N}$-valued random variable on $(\Omega, \mathcal{F}, \mathsf{P})$, independent of $(\Delta_i)_{i\geq 1}$ but having a possibly different distribution. Set

$$T_n := \Delta_0 + \Delta_1 + \ldots + \Delta_n.$$

The sequence $T := (T_n)_{n\in\mathbb{N}}$ is called a *renewal process*; $T_0 = \Delta_0$ is the *delay* of the process, and $\{T_n : n \geq 0\}$ is the set of *renewal times*. Observe that T is a Markov chain with respect to the filtration $\mathcal{F}_n := \sigma(\Delta_0, \ldots, \Delta_n)$, whose transition matrix has entries $A(i, j) := \mathsf{P}(\Delta_1 = j - i)$.

Let

$$p_k := \mathsf{P}(\Delta_1 = k)$$

for $k \in \mathbb{N}$. We say that T is *aperiodic* if $p_0 \neq 1$ and $\{k \geq 1 : p_k > 0\}$ is an aperiodic set as defined in Sect. 2.4. We say that T is L^p if Δ_1 is in L^p, i.e., $\sum_{k\in\mathbb{N}} k^p p_k < \infty$.

To fix ideas, one can imagine that a certain device breaks down and is replaced by a generic device at times $T_0, T_1, \ldots$. The lifespan of the initial device is distributed as Δ_0 and the lifespans of the replacement devices are distributed as Δ_1.

From now on we shall assume that T is aperiodic. For all $n \in \mathbb{N}$, let

$$\varsigma_n := \min\{k \geq 0 : T_k \geq n\}.$$

Then $\varsigma_n < \infty$ P-almost surely so that

$$X_n := T_{\varsigma_n} - n$$

is well-defined. A key observation is the following:

The set of renewal times for T equals the zero set of (X_n),

i.e.,

$$\{T_n : n \in \mathbb{N}\} = \{n \in \mathbb{N} : X_n = 0\}.$$

It is easily checked that with respect to the filtration $\{\mathcal{F}_{\varsigma_n}\}$, (X_n) is a Markov chain on $\mathbb{N}$ whose transition matrix is given by

$$P(k, k-1) = 1 \text{ for } k \geq 1,$$

$$P(0, k) = \frac{p_{k+1}}{1 - p_0} \text{ for } k \in \mathbb{N},$$

and

$$P(k, l) = 0 \text{ for } k \geq 1, l \neq k - 1.$$

Let $K := \sup\{k \geq 1 : p_k > 0\} \in \mathbb{N}^* \cup \{\infty\}$ and $M := \{0, \ldots, K-1\}$ (with the convention that $M = \mathbb{N}$ if $K = \infty$). Then $X_n \in M$ for n large enough (precisely $n \geq (X_0 - K + 1)^+$). On M, the chain (X_n) is irreducible, recurrent, and aperiodic (by aperiodicity of T).

Exercise 2.30 Verify the claims made about (X_n). In particular, show that (X_n) is a Markov chain with the transition matrix given above, and that (X_n) restricted to M is irreducible, recurrent, and aperiodic.

Let $\tau_0 = \inf\{n \geq 1 : X_n = 0\}$. Then

$$\mathbb{E}_0(\tau_0) = \sum_{k \geq 0}(1 + k)P(0, k) = \frac{\mathsf{E}(\Delta_1)}{1 - p_0} = \mathsf{E}(\Delta_1 | \Delta_1 > 0) \in (0, \infty],$$

where the expectation of a random variable X conditional on an event A of positive probability is defined as $\mathsf{E}(X|A) := \mathsf{E}(X\mathbf{1}_A)/\mathsf{P}(A)$. The equation $\mathbb{E}_0(\tau_0) = \mathsf{E}(\Delta_1)/(1 - p_0)$ implies that (X_n) is positive recurrent if and only if T is L^1.

Exercise 2.31 Assume that (X_n), restricted to M, is positive recurrent. Express the unique invariant probability measure for the transition matrix P in terms of the p_k's.

As a consequence of Theorem 2.27, we obtain the following classical renewal theorem.

Theorem 2.32 *Assume that T is aperiodic. Then*

$$\lim_{k \to \infty} \sum_{n=0}^{\infty} P(T_n = k) = \frac{1}{E(\Delta_1)},$$

with the convention that the right-hand side is zero if $E(\Delta_1) = \infty$.

Proof Let $N_k := \sum_{n \geq 0} 1_{\{T_n = k\}}$. Then

$$N_k = 1_{\{X_k = 0\}} \left(1 + \sum_{i \geq 1} 1_{\{T_i' = 0\}}\right),$$

where

$$T_i' := \Delta_{\varsigma_k + 1} + \ldots + \Delta_{\varsigma_k + i}.$$

Thus $E(N_k) = E(E(N_k | \mathcal{F}_{\varsigma_k})) = P(X_k = 0)\frac{1}{1 - p_0}$, and by Theorems 2.27 and 2.6,

$$\lim_{k \to \infty} P(X_k = 0) = \frac{1}{\mathbb{E}_0(\tau_0)}.$$

This proves the result. $\square$

2.6.1 Coupling of Renewal Processes

Suppose that T is L^1, and let $\tilde{T}$ be another aperiodic L^1-renewal process independent of T with

$$\tilde{T}_n = \tilde{\Delta}_0 + \tilde{\Delta}_1 + \ldots + \tilde{\Delta}_n.$$

The distribution of $(\tilde{\Delta}_i)_{i \geq 0}$ may be different from the one of $(\Delta_i)_{i \geq 0}$. We are interested in the first time $\tau > 0$ that is a renewal time for both T and $\tilde{T}$. Equivalently, with $\tilde{X}_n$ defined in analogy to X_n,

$$\tau := \inf\{n \geq 1 : X_n = \tilde{X}_n = 0\}.$$

We know that (X_n) is absorbed by M in finite time and that it is aperiodic and positive recurrent on M. Hence, $(X_n, \tilde{X}_n)$ is absorbed by $M \times \tilde{M}$ in finite time ($\tilde{M}$ defined in analogy to M) and, by Corollary 2.23, it is positive recurrent on $M \times \tilde{M}$. In particular,

$$\mathsf{P}(\tau < \infty) = \mathbb{P}_{\alpha \otimes \tilde{\alpha}}(\tau_{0,0} < \infty) = 1, \tag{2.5}$$

where α (respectively $\tilde{\alpha}$) denotes the law of Δ_0 (respectively $\tilde{\Delta}_0$). It turns out that whenever $\Delta_0, \tilde{\Delta}_0$ and $\Delta_1, \tilde{\Delta}_1$ are in L^p for some $p \geq 1$, the same is true for τ. A proof of this fact can be found for instance in Lindvall's book [47] and goes back to Pitman's seminal paper [55]. We provide here a short proof (different from Lindvall's) based on Proposition 2.18 and Theorem 2.19.

Theorem 2.33 *Suppose T and $\tilde{T}$ are aperiodic and in L^p for some $p \geq 1$. Then there exists a constant $c > 0$, independent of the distributions of Δ_0 and $\tilde{\Delta}_0$, such that $\mathsf{E}(\tau^p) \leq c(1 + \mathsf{E}(\Delta_0^p) + \mathsf{E}(\tilde{\Delta}_0^p))$.*

Proof Let $Q := P \otimes \tilde{P}$ denote the kernel of $(X_n, \tilde{X}_n)$. Let V be the function defined on $\mathbb{N} \times \mathbb{N}$ by $V(i, j) = \max(i, j) + 1$. One has

$$QV(i, j) - V(i, j) = -1 \text{ for } i \neq 0, j \neq 0,$$

and (by integrability of Δ_1 and dominated convergence)

$$\lim_{j \to \infty} QV(0, j) - V(0, j) = \lim_{j \to \infty} \mathsf{E}(\max(\Delta_1 - j - 1, -1)|\Delta_1 > 0) = -1.$$

Similarly, $\lim_{i \to \infty} QV(i, 0) - V(i, 0) = -2$. Condition (b) of Theorem 2.19 is then satisfied for the Markov process $(X_n, \tilde{X}_n)$ on $\mathbb{N} \times \mathbb{N}$, with $C = \{(i, j) \in \mathbb{N} \times \mathbb{N} : V \leq R\}$ and R large enough. Condition (b') is easily seen to be satisfied as well because Δ_1 and $\tilde{\Delta}_1$ are in L^p. Therefore, there is $c > 0$ such that for all $(i, j) \in \mathbb{N} \times \mathbb{N}$,

$$\mathbb{E}_{i,j}(\tau_{0,0}^p) \leq 2^{p-1}(\mathbb{E}_{i,j}(\tau_C^p) + \max_{(i,j) \in C} \mathbb{E}_{i,j}(\tau_{0,0}^p)) \leq c(1 + \max(i, j)^p). \tag{2.6}$$

Here, the first inequality follows from the strong Markov property and inequality $\tau_{0,0} \leq \tau_C + \tau_{0,0} \circ \Theta_{\tau_C}$. The second inequality follows from Theorem 2.19 (iii) and Proposition 2.18. Note that while $(X, \tilde{X})$ is not necessarily irreducible on $\mathbb{N} \times \mathbb{N}$ and thus a key assumption of Proposition 2.18 is not satisfied, the proof still goes through because any point $(i, j) \in \mathbb{N} \times \mathbb{N}$ leads to $(0, 0)$. Integrating the inequality in (2.6) with respect to $\alpha \otimes \tilde{\alpha}$, the law of $(\Delta_0, \tilde{\Delta}_0) = (X_0, \tilde{X}_0)$, gives the result. $\qquad \square$

Theorem 2.34 *Suppose T and $\tilde{T}$ are aperiodic and*

$$\mathsf{E}(e^{\lambda_0 \Delta_1}) + \mathsf{E}(e^{\lambda_0 \tilde{\Delta}_1}) < \infty$$

for some $\lambda_0 > 0$. Then there exist $0 < \lambda \le \lambda_0$ and $c > 0$ such that

$$\mathsf{E}(e^{\lambda \tau}) \le c(1 + \mathsf{E}(e^{\lambda_0 \Delta_0}) + \mathsf{E}(e^{\lambda_0 \tilde{\Delta}_0})).$$

Proof The proof is similar to the proof of Theorem 2.33. Set $V(i, j) := e^{\lambda_0 i} + e^{\lambda_0 j}$. Then $QV(i, j) \le e^{-\lambda_0} V(i, j) + \kappa$ with

$$\kappa := \mathsf{E}(e^{\lambda_0 \Delta_1} | \Delta_1 > 0) + \mathsf{E}(e^{\lambda_0 \tilde{\Delta}_1} | \tilde{\Delta}_1 > 0).$$

Condition (c) of Theorem 2.19 is then satisfied for any $0 < \lambda < 1 - e^{-\lambda_0}$ and $C = \{(i, j) \in \mathbb{N} \times \mathbb{N} : V(i, j) \le R\}$ with R sufficiently large given the choice of λ (see also Exercise 2.20). Then, relying on $\tau_{0,0} \le \tau_C + \tau_{0,0} \circ \Theta_{\tau_C}$, the strong Markov property, Theorem 2.19 (iv), and Proposition 2.18, we obtain

$$\mathbb{E}_{i,j}(e^{\lambda \tau_{0,0}}) \le c(1 + V(i, j)), \quad \forall (i, j) \in \mathbb{N} \times \mathbb{N}$$

for some $c > 0$ and some $\lambda \in (0, 1 - e^{-\lambda_0})$. Integrating this inequality with respect to the law of $(\Delta_0, \tilde{\Delta}_0)$ gives the desired result.   □

2.7 Convergence Rates for Positive Recurrent Chains

We revisit here the ergodic theorems from Sect. 2.5, Theorems 2.27 and 2.28, with the help of Theorems 2.33 and 2.34.

Let M be countable and let $(X_n, Y_n)_{n \ge 0}$ be the canonical chain on $(M \times M)^{\mathbb{N}}$. Let P be an irreducible, aperiodic, and positive recurrent kernel on M. If π denotes the invariant probability measure of P, we have seen in the proofs of Theorems 2.27 and 2.28 that for every probability measure μ on M and every $x^* \in M$,

$$\sup_{x \in M} |\mu P^n(x) - \pi(x)| \le \mathbb{P}_{\mu \otimes \pi}(\tau_{(x^*, x^*)} > n),$$

where $\mathbb{P}_{\mu \otimes \pi}$ is the Markov measure with kernel $P \otimes P$ and initial distribution $\mu \otimes \pi$, and where $\tau_{(x^*, x^*)} = \inf\{n \ge 1 : X_n = Y_n = x^*\}$.

Let $(\tau_{x^*}^{(n)})$ (respectively $(\tilde{\tau}_{x^*}^{(n)})$) denote the successive hitting times of x^* by (X_n) (respectively (Y_n)). Then, for any probability measures α, β on M, the processes $T := (\tau_{x^*}^{(n+1)})_{n \ge 0}$ and $\tilde{T} := (\tilde{\tau}_{x^*}^{(n+1)})_{n \ge 0}$ living on the probability space $((M \times M)^{\mathbb{N}}, \mathcal{B}((M \times M)^{\mathbb{N}}), \mathbb{P}_{\alpha \otimes \beta})$ are two independent renewal processes and $\tau_{(x^*, x^*)}$ is nothing but the first common renewal time for T and $\tilde{T}$.

The Markov inequality, Theorems 2.33, 2.34, and Proposition 2.10 lead to the following result.

Theorem 2.35 *Let P be irreducible, aperiodic, and positive recurrent, with invariant probability measure π. Let $x^* \in M$.*

(i) *If $\mathbb{E}_{x^*}(\tau_{x^*}^p) < \infty$ for some $p \geq 2$, then there exists $c \geq 0$ such that for every probability measure μ on M and for every $n \in \mathbb{N}^*$,*

$$\sup_{x \in M} |\mu P^n(x) - \pi(x)| \leq \frac{1}{n^{p-1}} c(1 + \mathbb{E}_\mu(\tau_{x^*}^{p-1})).$$

(ii) *If $\mathbb{E}_{x^*}(e^{\lambda_0 \tau_{x^*}}) < \infty$ for some $\lambda_0 > 0$, then there exist $0 < \lambda < \lambda_0$ and $c \geq 0$ such that for every probability measure μ on M and for every $n \in \mathbb{N}$,*

$$\sup_{x \in M} |\mu P^n(x) - \pi(x)| \leq e^{-\lambda n} c(1 + \mathbb{E}_\mu(e^{\lambda_0 \tau_{x^*}})).$$

Combined with Theorem 2.19, Proposition 2.18, and the strong Markov property, we recover and extend Theorem 2.28.

Corollary 2.36 *Let P be irreducible, aperiodic, and positive recurrent, with invariant probability measure π. Let $V : M \to [1, \infty)$ and let $C \subset M$ be as in Theorem 2.19 ((b') or (c)) with C finite. Then*

(i) *Under condition (b') of Theorem 2.19 for $p \geq 2$, there is $c \geq 0$ such that for every probability measure μ on M and for every $n \in \mathbb{N}^*$,*

$$\sup_{x \in M} |\mu P^n(x) - \pi(x)| \leq \frac{1}{n^{p-1}} c(1 + \mu V^p);$$

(ii) *Under condition (c) of Theorem 2.19, there are $c, \lambda > 0$ such that for every probability measure μ on M and for every $n \in \mathbb{N}$,*

$$\sup_{x \in M} |\mu P^n(x) - \pi(x)| \leq e^{-\lambda n} c(1 + \mu V).$$

Notes

The book by Aldous and Fill [1] contains numerous interesting identities for the mean hitting times ($\mathbb{E}_x(\tau_y)$), the occupation times ($\mathbb{E}_x(N_y)$) and their relation to the rate of convergence. Convergence rates for finite Markov chains, in terms of the geometry of the chain, are thoroughly investigated in the monograph by Saloff-Coste [62] and the book by Levin, Peres, and Wilmer [46]. A nice extension of Chung's theorem can be found in the recent paper [3]. The coupling method leading to the convergence rate Theorem 2.35 goes back to Pitman [55] (see also Lindvall's book [47]).

Chapter 3
Random Dynamical Systems

Whether it is on countable or non-countable state spaces, numerous examples of Markov chains are given by *random dynamical systems* (also called *random iterative systems*). These are systems defined by a recursion of the form $X_{n+1} = F_{\theta_{n+1}}(X_n)$ where (θ_n) is a sequence of independent identically distributed random variables. This short chapter discusses their basic properties and the question of the representation of a general (respectively Feller) Markov chain by a random dynamical system.

3.1 General Definitions

Let $(\Theta, \mathcal{A}, m)$ be a probability space,

$$F : \Theta \times M \to M$$

$$(\theta, x) \mapsto F_\theta(x),$$

a measurable map, and $(\theta_n)_{n \geq 1}$ a sequence of independent identically distributed (i.i.d.) Θ-valued random variables having law m. Consider an M-valued process recursively defined by

$$X_{n+1} := F_{\theta_{n+1}}(X_n) \tag{3.1}$$

for some given random variable X_0.

© The Author(s), under exclusive license to Springer Nature Switzerland AG 2022
M. Benaïm, T. Hurth, *Markov Chains on Metric Spaces*, Universitext,
https://doi.org/10.1007/978-3-031-11822-7_3

Proposition 3.1 *Assume that X_0 is a random variable independent of (θ_n). Then (X_n) is a Markov chain on M whose Markov kernel is given by*

$$P(x, G) = m(\theta \in \Theta : F_\theta(x) \in G). \tag{3.2}$$

If furthermore F_θ is continuous for m-almost every θ, then P is Feller.

Proof The proof follows (almost) directly from the definitions. Measurability of $x \mapsto P(x, G)$ is a by-product of Fubini's theorem since $P(x, G) = \int_\Theta 1_G \circ F_\theta(x)\, m(d\theta)$. The Feller property follows from continuity under the integral sign.
□

The kernel P defined by (3.2) is called the *Markov kernel induced by* (F, m). The sequence of random maps (F^n) defined by

$$F^n := F_{\theta_n} \circ F_{\theta_{n-1}} \circ \ldots \circ F_{\theta_1}$$

is called the *random dynamical system* (RDS) *induced by* (F, m).

Note that, by Chapman-Kolmogorov, the law of $F^n(x)$ is determined by P ($F^n(x)$ has law $P^n(x, \cdot)$) but, as shown by the next example, P is not sufficient to characterize the law of F^n.

Example 3.2 This example is due to Kifer [43]. Let $M = S^1 = \{z \in \mathbb{C} : |z| = 1\}$ be the unit circle, $\Theta = [0, 1]$, and $m(d\theta) = d\theta$ the uniform Lebesgue measure. Let $f : S^1 \to S^1$ be any, say continuous, map and $F_\theta(z) = e^{2i\pi\theta} f(z)$. Then $P(z, \cdot)$ is the uniform measure on S^1 for every $z \in S^1$, but the random dynamical system induced by (F, m) clearly depends on the choice of f. For instance, if $f(z) = z$, F^n preserves the distance between points, while for $f(z) = z^2$, F^n locally increases the distance exponentially.

Example 3.3 This example is due to Diaconis and Freedman [19]. Let $M = [0, 1]$ be the closed unit interval, and

$$P(x, dy) = \frac{1}{2x} 1_{[0,x]}(y)dy + \frac{1}{2(1-x)} 1_{[x,1]}(y)dy.$$

Here we adopt the convenient convention that $\frac{1_{[0,x]}(y)}{x}dy = \delta_0(dy)$ for $x = 0$ and $\frac{1_{[x,1]}(y)}{(1-x)}dy = \delta_1(dy)$ for $x = 1$. In words, if the chain is at x it moves to a point y randomly chosen in the right interval $[x, 1]$ (respectively left interval $[0, x]$) with probability $1/2$.

Let $F : (0, 1) \times [0, 1] \to [0, 1]$ be defined by

$$F_\theta(x) := 2\theta x 1_{\theta < 1/2} + [x + (2\theta - 1)(1 - x)]1_{\theta \geq 1/2}.$$

Then P is induced by (F, dx).

Exercise 3.4 (Additive Noise) Suppose $M = \mathbb{R}^n$ (or an abelian locally compact group), $\Theta = M$, $F : M \to M$,

$$F_\theta(x) = F(x) + \theta$$

and $m(d\theta) = h(\theta)d\theta$ with $h \in L^1(d\theta)$. Here $d\theta$ stands for the Lebesgue measure (or the Haar measure) on M. Let P denote the corresponding Markov kernel given by (3.2).

Given $x \in M$, let $U_x : L^1(dx) \to L^1(dx)$ be the translation operator defined as $U_x(g)(y) := g(y - x)$. Show that for all $f \in B(M)$,

$$|Pf(x) - Pf(y)| \le \|f\|_\infty \|U_{F(x)}(h) - U_{F(y)}(h)\|_1.$$

Deduce that P is strong Feller whenever F is continuous. One can use (or better, prove) that for all $g \in L^1(dx)$, $x \in M \mapsto U_x(g) \in L^1(dx)$ is continuous.

3.2 Representation of Markov Chains by RDS

Proposition 3.1 shows that every RDS defines a Markov chain. Here we briefly discuss the converse problem and consider the question of representing a Markov chain by a suitable RDS.

A *transformation space* is a set of maps $f : M \to M$ closed under composition. Let **T** be a transformation space and P a Markov kernel on M.

We say that P can be represented by **T** if there exists a probability space $(\Theta, \mathcal{A}, m)$ and a measurable map $F : \Theta \times M \to M$ such that

 (i) $F_\theta \in \mathbf{T}$ for all $\theta \in \Theta$;
 (ii) P is induced by (F, m).

Recall that a separable metric space M is called *Polish* if it is complete. The following result is folklore.

Theorem 3.5 *If M is a Borel subset of a Polish space, then any Markov kernel on M can be represented by a space $\mathbf{T}$ of measurable maps with $(\Theta, \mathcal{A}, m) = ((0, 1), \mathcal{B}((0, 1)), \lambda)$ and λ the Lebesgue measure on $(0, 1)$.*

Proof When M is a Borel subset of $\mathbb{R}$, the proof is constructive and makes F explicit. Indeed, let G_x be the cumulative distribution function of $P(x, .)$, i.e.,

$$G_x(t) = P(x, (-\infty, t]).$$

For all $\theta \in (0, 1)$ and $x \in M$, set

$$F_\theta(x) := G_x^{-1}(\theta),$$

where $G_x^{-1} : (0, 1) \to \mathbb{R}$, the generalized inverse of G_x, is defined as

$$G_x^{-1}(u) := \inf\{t \in \mathbb{R} : G_x(t) \geq u\}.$$

Then

$$\lambda(\theta \in (0, 1) : F_\theta(x) \leq t) = \lambda(\theta \in (0, 1) : \theta \leq G_x(t)) = G_x(t).$$

The proof in the general case follows from the following abstract result of measure theory: Every Borel subset M of a Polish space is *isomorphic* to a Borel subset of $[0, 1]$. That is, there exists a Borel set $\tilde{M} \subset [0, 1]$ and a bi-measurable bijection $\Psi :$ $M \to \tilde{M}$ (meaning that both Ψ and its inverse are Borel measurable). Chapter 13 of Dudley's book [21] contains a detailed proof of this result. Exercise 4.11 treats the particular case where M is compact or locally compact.

Given such a Ψ and a Markov kernel P on M, let $\tilde{P}$ be the Markov kernel on $\tilde{M}$ defined as $\tilde{P}(x, A) := P(\Psi^{-1}(x), \Psi^{-1}(A))$. Then $\tilde{P}$ is induced by $(\tilde{F}, \lambda)$ for some measurable $\tilde{F} : (0, 1) \times \tilde{M} \to \tilde{M}$ so that P is induced by (F, λ) with $F_\theta(x) = \Psi^{-1}(\tilde{F}_\theta(\Psi(x)))$. $\qquad\square$

Blumenthal and Corson [12] prove the following result (see also Kifer [43], Theorem 1.2).

Theorem 3.6 ([12]) *Let M be a connected and locally connected compact metric space. Let P be a Feller Markov kernel such that $P(x, \cdot)$ has full support for all $x \in M$, i.e., for all $x \in M$ and for every closed set F strictly contained in M, we have $P(x, F) < 1$. Then P may be represented by $\mathbf{T} = C^0(M, M)$ (the space of continuous maps $f : M \to M$).*

The question of representation by smooth maps has been considered by Quas [58]. Before stating Quas's theorem, we state a result due to Jürgen Moser from which it will be deduced.

Let M be a smooth (C^∞) compact orientable Riemannian manifold without boundary, with normalized Riemannian probability measure λ. If $\rho : M \to \mathbb{R}_+$ is a C^1-density on M and $\Phi : M \to M$ a C^1-diffeomorphism, we let $\Phi^*\rho$ denote the image of ρ by Φ, i.e.,

$$(\Phi^*\rho)(\Phi(x)) = \frac{\rho(x)}{|J\Phi(x)|},$$

where $J\Phi(x)$ is the Jacobian of Φ, i.e., the determinant of the derivative $D\Phi(x) :$ $T_x M \to T_{\Phi(x)}M$. In other words, if X is a random variable with density ρ, then $\Phi(X)$ is a random variable with density $\Phi^*\rho$.

In 1965, Moser [50], using the "homotopy trick" argument, proved part (i) of the following result in the C^∞ case. For every positive integer k and $0 \leq \alpha < 1$, we let $C^{k+\alpha}(M)$ denote the space of $C^{k+\alpha}$ (C^k with α-Hölder kth derivatives if $\alpha > 0$)

functions $h : M \to \mathbb{R}$ endowed with the $C^{k+\alpha}$-topology,

$$E_t^{k+\alpha} := \{h \in C^{k+\alpha}(M) : \int_M h(x)\lambda(dx) = t\},$$

and $D^{k+\alpha} := \{\rho \in E_1^{k+\alpha} : \rho(x) > 0 \ \forall x \in M\}$ the space of positive $C^{k+\alpha}$-densities. Plainly, $E_1^{k+\alpha}$ is a closed subset of $C^{k+\alpha}(M)$, which can be identified with the Banach space $E_0^{k+\alpha}$, and $D^{k+\alpha}$ is an open subset of $E_1^{k+\alpha}$.

Theorem 3.7 ([50]) *Let ρ_0 be a positive C^k-density for some $k \geq 1$. Then*

(i) *For any positive C^k-density ρ, there exists a C^k-diffeomorphism Φ_ρ on M with the property that*

$$\Phi_\rho^* \rho_0 = \rho;$$

(ii) *The C^k-diffeomorphism Φ_ρ from part (i) can be chosen in such a way that the mapping*

$$D^k \times M \to M,$$

$$(\rho, x) \mapsto \Phi_\rho(x)$$

is C^k.

Proof Let $\rho_t = \rho_0 + t(\rho - \rho_0)$ for $0 \leq t \leq 1$. We look for a family of diffeomorphisms $(\Phi_t)_{t\in[0,1]}$ such that $\Phi_t^* \rho_0 = \rho_t$ for all $t \in [0, 1]$. That is,

$$j(t, x)\rho_t(\Phi_t(x)) = \rho_0(x), \qquad (3.3)$$

where $j(t, x)$ is the Jacobian of Φ_t, evaluated at x. More precisely, we look for a family of vector fields $\{X_t\}_{t\in[0,1]}$ on M such that $\Phi_t(x)$ is the solution to the non-autonomous Cauchy problem

$$\frac{dy}{dt} = X_t(y)$$

with initial condition $y(0) = x$. Using Jacobi's formula for the derivative of the determinant of a matrix-valued function, one obtains that $j(t, x)$ solves

$$\frac{dj}{dt} = div(X_t)[\Phi_t(x)]j(t)$$

with initial condition $j(0, \cdot) \equiv 1$. Thus, taking the time derivative of (3.3) and setting $y := \Phi_t(x)$, $\eta := \rho_0 - \rho$ gives

$$div(X_t)(y)\rho_t(y) - \eta(y) + \langle \nabla \rho_t(y), X_t(y) \rangle_y = 0.$$

Hence

$$div(\rho_t X_t)(y) = \eta(y).$$

If one sets $X_t = \nabla U / \rho_t$, the problem reduces to finding a function $U : M \to \mathbb{R}$ such that

$$\Delta U = div(\nabla U) = \eta, \qquad\qquad (3.4)$$

where one should recall that $\eta = \rho_0 - \rho$.

Since

$$\int_M \eta(x)\, \lambda(dx) = 0,$$

(3.4) admits a solution, and we may define $\Delta^{-1}\eta$ as the particular solution

$$x \mapsto 2 \int_0^\infty Q_t \eta(x) dt,$$

where $Q_t \eta(x) := E(\eta(W_t)|W_0 = x)$ and W_t is a Brownian motion on M. Furthermore, by Schauder estimates (see, e.g., Chapter 6 in [30]) Δ^{-1} maps $E_0^{k-1+\alpha}(M)$ continuously into $C^{k+1+\alpha}(M)$ for every positive integer k and $0 < \alpha < 1$. This makes the vector field

$$X_t^\rho := \nabla U / \rho_t$$

a C^k-vector field. It also implies that the continuous mapping

$$[0, 1] \times D^k \times M \to TM,$$

$$(t, \rho, x) \mapsto X_t^\rho(x)$$

is C^k.

Let $t \mapsto \Phi_t(\rho, x)$ denote the solution to the Cauchy problem $\frac{dy}{dt} = X_t^\rho(y)$ with initial condition $\Phi_0(\rho, x) = x$. It then follows from standard results on differential equations that $x \mapsto \Phi_t(\rho, x)$ is a C^k-diffeomorphism for all $(t, \rho) \in [0, 1] \times D^k$, and that $(x, \rho) \mapsto \Phi_t(\rho, x)$ is C^k for all $t \in [0, 1]$. To conclude the proof, set $\Phi_\rho(x) := \Phi_1(\rho, x)$.   □

From Moser's theorem we deduce the following result proved by Quas [58] in the C^∞ case.

Corollary 3.8 ([58]) *Let P be a Markov kernel on M, a smooth compact orientable connected Riemannian manifold without boundary. Assume that for each $x \in M$, P_x has a C^k, $k \geq 1$, positive density ρ_x with respect to the Riemannian measure, and that $x \in M \mapsto \rho_x \in D^k$ is C^r, $r \geq 0$. Then P may be represented by $\mathbf{T} = C^r(M, M)$.*

Proof Let $\rho_0 = \rho_{x_0}$ for some $x_0 \in M$ and let $\Psi_x = \Phi_{\rho_x}$ denote the C^k-diffeomorphism produced by Moser's Theorem (Theorem 3.7). Then

$$P(x, G) = P(x_0, \Psi_x^{-1}(G)).$$

Let $\mathbf{T} = C^r(M, M)$ and let $f_y \in \mathbf{T}$ be defined by $f_y(x) := \Psi_x(y)$. Then

$$P(x, G) = m(f \in \mathbf{T} : f(x) \in G),$$

where m is the image of P_{x_0} by the mapping $y \in M \mapsto f_y \in \mathbf{T}$. □

Exercise 3.9 (Bernoulli Convolutions) Bernoulli convolutions are very simple, still fascinating, examples of random dynamical systems.

Let $0 < a < 1$ and let (X_n) be the sequence of real-valued random variables recursively defined by

$$X_{n+1} = aX_n + \theta_{n+1},$$

where (θ_n) is a sequence of i.i.d. random variables taking values in $\{-1, 1\}$, independent of X_0, and having uniform distribution $m = \frac{\delta_{-1} + \delta_1}{2}$.

Set $Y_n = \sum_{i=0}^{n-1} a^i \theta_{i+1}$ and let

$$Y = \lim_{n \to \infty} Y_n = \sum_{i \geq 0} a^i \theta_{i+1}.$$

Throughout, we let μ_a denote the law of Y and F_a its cumulative distribution function (cdf) defined as $F_a(t) = \mu_a((-\infty, t])$.

(i) Show that $X_n - a^n X_0$ and Y_n have the same law and deduce that (X_n) *converges in law* to μ_a, i.e.,

$$\lim_{n \to \infty} \mathsf{E}(f(X_n)) = \mu_a f$$

for all $f \in C_b(\mathbb{R})$. Convergence in law will be further discussed in Sect. 4.1 of Chap. 4.

(ii) Show that F_a is the unique cdf solution to the functional equation

$$F(t) = \frac{1}{2}\left[F\left(\frac{t-1}{a}\right) + F\left(\frac{t+1}{a}\right) \right].$$

(iii) Show that F_a is continuous.
(iv) (Law of pure types) Recall that μ_a is called *absolutely continuous* (with respect to Lebesgue measure) if every Borel set having zero Lebesgue measure has zero μ_a-measure. By the Radon–Nikodym theorem, this amounts to

$$F_a(t) = \int_{-\infty}^{t} f_a(u)\,du$$

for some nonnegative function $f_a \in L^1(\mathbb{R})$. The measure μ_a is called *singular* if $\mu_a(N) = 1$ for some Borel set N having zero Lebesgue measure. Show that μ_a is either absolutely continuous or singular (compare with Lemma 4.26 in Chap. 4).
(v) (Devil's staircase) The *topological support* of μ_a is the set of $t \in \mathbb{R}$ such that $\mu_a(I) > 0$ for every open interval I containing t. Equivalently, this is the set of $t \in \mathbb{R}$ at which F_a strictly increases.

Suppose $a < \frac{1}{2}$. Show that the support of μ_a is a Cantor set having zero Lebesgue measure. In this case F_a is a *Devil's staircase*: a continuous function increasing from 0 to 1 but almost everywhere nonincreasing.
(vi) Show that $\mu_{1/2}$ is the uniform distribution over $[-2, 2]$.
(vii) Show that for $a > \frac{1}{2}$, the support of μ_a is the interval $[-\frac{1}{1-a}, \frac{1}{1-a}]$.

Remark 3.10 The study of Bernoulli convolutions has a long history. It started around 1930 with the work of Wintner and his collaborators Jessen and Kershner (see, e.g., [53] for a comprehensive bibliography). As seen in the previous exercise, when $a > \frac{1}{2}$, F_a is continuous and strictly increasing on $[-\frac{1}{1-a}, \frac{1}{1-a}]$. Wintner proved that it is C^{k-1} for $a = 2^{-1/k}$ and $k \geq 2$, but Erdös [25] in 1939 proved that whenever $\frac{1}{a}$ is a *Pisot number*, then μ_a is singular! A Pisot number is a real algebraic integer (i.e., the root of a unitary polynomial having integer coefficients) whose conjugates (i.e., the other roots of the polynomial) have modulus < 1. For instance, the golden number $g = \frac{1+\sqrt{5}}{2}$ is a Pisot number as the root of the polynomial $X^2 - X - 1$.

After Erdös, the question of describing the set of $a > \frac{1}{2}$ for which μ_a is absolutely continuous has challenged the community. In 1995 Solomyak [64] (see also the beautiful short proof by Solomyak and Peres [54]) proved the remarkable result that for almost all $a > \frac{1}{2}$, μ_a is absolutely continuous.

Exercise 3.11 (The Propp and Wilson Algorithm) The representation of a Markov chain by a RDS can obviously be used to simulate trajectories of a given finite Markov chain. More surprisingly it can also serve to sample *exactly* and *in finite time* the invariant probability measure of a positive recurrent finite chain. This is the Propp and Wilson algorithm introduced by J. Propp and D. Wilson [57] in 1996.

Let M be a finite set and let (F^n) be a RDS on M. Recall that this means that

$$F^n = F_{\theta_n} \circ \ldots \circ F_{\theta_1},$$

where (θ_i) is a sequence of i.i.d. random variables on some probability space $(\Theta, \mathcal{A}, m)$ and $\Theta \times M \ni (\theta, x) \mapsto F_\theta(x)$ is a measurable map.

Associated to F^n is the *right product*

$$R^n = F_{\theta_1} \circ \ldots \circ F_{\theta_n}.$$

A map $f : M \to M$ is called *constant* if $f(x) = f(y)$ for all $x, y \in M$. We let Cst denote the set of such maps, and

$$T_c = \min\{n \geq 0 : R^n \in \mathsf{Cst}\}.$$

(i) Show that R^n and F^n have the same distribution.
(ii) Suppose that T_c is almost surely finite. Let $Z = R^{T_c}(x)$ (which is independent of x). Show that for all $n \geq T_c$ and $y \in M$, $R^n(y) = Z$. Deduce that the law of Z is the unique invariant probability measure of the chain induced by (F^n).
(iii) Suppose that for some $\alpha > 0$, $m(\{\theta \in \Theta : F_\theta \in \mathsf{Cst}\}) \geq \alpha$. Show that T_c has a geometric tail, and is therefore almost surely finite.
(iv) Suppose, more generally, that for some $\alpha > 0$ and every subset $A \subset M$ having cardinality $|A| \geq 2$,

$$m(\{\theta \in \Theta : |F_\theta(A)| < |A|\}) \geq \alpha.$$

Show that T_c has a geometric tail and is therefore almost surely finite.
(v) Suppose now that P is a Markov transition matrix on M having positive entries. Show that it is always possible to represent it by a RDS such that the condition assumed in question (iii) is satisfied. Explain how this can be used to produce an algorithm which samples the invariant probability measure of P in finite time.
(vi) Let $M = \{0, 1\}$ and let P be the Markov transition matrix defined by $P(x, y) = \frac{1}{2}$. Let $\Theta = \{0, 1\}$, $m = \frac{1}{2}(\delta_0 + \delta_1)$, and $F_\theta(x) = \theta x + (1 - \theta)(1 - x)$. Show that the Markov kernel P is represented by (F, m) but that $T_c = \infty$ almost surely.

Notes

The proof of Erdös's theorem on Bernoulli convolutions (see Remark 3.10) as well as numerous illustrating simulations can be found in the first chapter of [7]. For (much) more on Bernoulli convolutions we recommend the survey papers [53] and [67]. The book [46] contains a full chapter on the Propp and Wilson algorithm including many examples of applications.

Chapter 4
Invariant and Ergodic Probability Measures

Invariant and ergodic probability measures are at the heart of the (ergodic) theory of Markov chains. This chapter starts with a brief summary of weak convergence theory, which we will use throughout the book. We then define invariant measures and show that limit points of the empirical occupation measures of a Feller chain are invariant probability measures. The rest of this chapter is devoted to ergodicity. Basic properties of ergodic measures are established and unique ergodicity of "random contractions" is proved. An entire section is devoted to the fundamental results of deterministic ergodic theory, namely the Poincaré recurrence theorem, the Birkhoff ergodic theorem, and the ergodic decomposition theorem. In another section, we present the Markovian versions of these results. In the final section, it is shown how the theory can be adapted to deal with continuous-time processes.

4.1 Weak Convergence of Probability Measures

Let $\mathcal{P}(M)$ denote the set of probability measures on $(M, \mathcal{B}(M))$. A sequence $\{\mu_n\} \subset \mathcal{P}(M)$ is said to *converge weakly* to $\mu \in \mathcal{P}(M)$, written

$$\mu_n \Rightarrow \mu,$$

provided

$$\lim_{n \to \infty} \mu_n f = \mu f$$

for all $f \in C_b(M)$. The following theorem, known as Portmanteau theorem, gives equivalent conditions for weak convergence. Note that this theorem is true in any metric space (without assumption of separability or completeness).

© The Author(s), under exclusive license to Springer Nature Switzerland AG 2022
M. Benaïm, T. Hurth, *Markov Chains on Metric Spaces*, Universitext,
https://doi.org/10.1007/978-3-031-11822-7_4

Let $U_b(M) \subset C_b(M)$ (resp. $L_b(M) \subset U_b(M)$) denote the set of bounded and uniformly continuous (resp. bounded and Lipschitz) mappings $f : M \to \mathbb{R}$.

Theorem 4.1 (Portmanteau Theorem) *Let $\{\mu_n\} \subset \mathcal{P}(M)$ and $\mu \in \mathcal{P}(M)$. The following conditions are equivalent:*

(a) $\mu_n \Rightarrow \mu$;
(b) $\mu_n f \to \mu f$ *for all* $f \in U_b(M)$;
(c) $\mu_n f \to \mu f$ *for all* $f \in L_b(M)$;
(d) $\limsup_{n\to\infty} \mu_n(F) \le \mu(F)$ *for all closed sets* $F \subset M$;
(e) $\liminf_{n\to\infty} \mu_n(O) \ge \mu(O)$ *for all open sets* $O \subset M$;
(f) $\lim_{n\to\infty} \mu_n(A) = \mu(A)$ *for all* $A \in \mathcal{B}(M)$ *such that* $\mu(\partial A) = 0$, *where* $\partial A := \overline{A} \setminus int(A)$ *denotes the boundary of* A.

Proof $(a) \Rightarrow (b) \Rightarrow (c)$ is clear and $(d) \Leftrightarrow (e)$ holds by set complementation.

Assume (c). Let F be a closed set, $\varepsilon > 0$, and $f_\varepsilon(x) := (1 - \frac{d(x,F)}{\varepsilon})^+$, where $d(x, F) := \inf_{y \in F} d(x, y)$. Then $1 \ge f_\varepsilon \ge \mathbf{1}_F$ and $f_\varepsilon \in L_b(M)$. Thus, $\limsup \mu_n(F) \le \limsup \mu_n f_\varepsilon = \mu f_\varepsilon$ and, by dominated convergence, $\mu f_\varepsilon \to \mu(F)$ as $\varepsilon \to 0$. This proves that $(c) \Rightarrow (d)$.

Assume (d) (and thus also (e)). Let $A \in \mathcal{B}(M)$ be such that $\mu(\partial A) = 0$. Let F be the closure of A and O its interior. Then $\mu(F) = \mu(O)$ and, by (d) and (e), $\liminf \mu_n(A) \ge \liminf \mu_n(O) \ge \mu(O)$ and $\limsup \mu_n(A) \le \limsup \mu_n(F) \le \mu(F)$. This proves that $(d), (e) \Rightarrow (f)$.

It remains to show that $(f) \Rightarrow (a)$. Assume (f) and let $f \in C_b(M)$. Replacing f by $f+c$ for some $c > 0$ if necessary, we can assume that $f \ge 0$. For all $a \ge 0$, the set $\{f > a\}$ is open and its boundary is contained in $\{f = a\}$. Furthermore, the set of $a \ge 0$ such that $\mu(\{f = a\}) > 0$ is at most countable (as the set of discontinuity points of the cumulative distribution function $a \mapsto \mu(\{f \le a\})$). Thus, by Fubini's theorem, (f), and dominated convergence, $\mu_n f = \int_0^{\|f\|_\infty} \mu_n(f > a)da \to \int_0^{\|f\|_\infty} \mu(f > a)da = \mu f$. $\qquad\square$

The following corollary is often useful.

Corollary 4.2 *Let $f \in B(M)$ and let D_f denote the set of discontinuities of f. If $\mu_n \Rightarrow \mu$ and $\mu(D_f) = 0$, then $\mu_n f \to \mu f$.*

Proof Let $\mu_n^f := \mu_n(f^{-1}(\cdot))$ be the image measure of μ_n by f. It suffices to show that $\mu_n^f \Rightarrow \mu^f$. Indeed, let $g(t) := t$ for $|t| \le \|f\|_\infty$, and $g(t) := \text{sign}(t)\|f\|_\infty$ for $|t| > \|f\|_\infty$. Then $\mu_n^f g = \mu_n f$ and $\mu^f g = \mu f$. To prove that $\mu_n^f \Rightarrow \mu^f$, we rely on assertion (d) of the Portmanteau theorem. Let F be a closed subset of $\mathbb{R}$. Then $\limsup \mu_n^f(F) \le \limsup \mu_n(\overline{f^{-1}(F)}) \le \mu(\overline{f^{-1}(F)})$ because $\mu_n \Rightarrow \mu$. Now, $\overline{f^{-1}(F)} \subset D_f \cup f^{-1}(F)$ so that $\mu(\overline{f^{-1}(F)}) = \mu(f^{-1}(F)) = \mu^f(F)$. $\qquad\square$

Exercise 4.3 For $\varepsilon, \delta > 0$ let $A_{\varepsilon,\delta}$ be the set of $x \in M$ such that $|f(y) - f(z)| \ge \varepsilon$ for some $y, z \in B(x, \delta)$. Show that $D_f = \bigcup_{n\in\mathbb{N}^*} \bigcap_{m\in\mathbb{N}^*} A_{1/n,1/m}$ and that D_f is measurable (even if f is not).

Exercise 4.4 Let P be a Markov kernel on a metric space M. Show that P is Feller if and only if the map $\varphi : M \to \mathcal{P}(M)$, $x \mapsto P(x, \cdot)$ is continuous (where $\mathcal{P}(M)$ is equipped with the topology of weak convergence).

The space $\mathcal{P}(M)$ equipped with the topology of weak convergence is actually a metric space, as shown by the next proposition.

Proposition 4.5 *There exists a countable family* $\{f_n\}_{n \geq 0} \subset C_b(M)$ *such that*

$$D(\mu, \nu) := \sum_{n \geq 0} \frac{1}{2^n} \min(|\mu f_n - \nu f_n|, 1)$$

is a distance on $\mathcal{P}(M)$ *whose induced topology is the topology of weak convergence. That is,* $\mu_n \Rightarrow \mu$ *if and only if* $D(\mu_n, \mu) \to 0$.

Remark 4.6 Unless when M is compact, the family $\{f_n\}_{n \geq 0}$ is not dense in $C_b(M)$ (see Exercise 4.10).

Proof If M is compact, $C_b(M)$ is separable (see Exercise 4.9) and it suffices to choose a dense sequence $\{f_n\} \subset C_b(M)$. If M is not compact, $C_b(M)$ is no longer separable (see Exercise 4.10), but we shall prove that there exists a metric $\tilde{d}$ on M, topologically equivalent to d, making M homeomorphic to a subset of a compact metric space. It will then follow that $U_b(M, \tilde{d})$, the space of bounded uniformly continuous functions on $(M, \tilde{d})$, is separable. (Here one should recall that two topologically equivalent metrics may yield distinct sets of uniformly continuous functions.)

Replacing d by $\frac{d}{1+d}$ (which remains a distance on M inducing the same topology as d), we can assume that $d \leq 1$. Let $\{a_n\}_{n \geq 0} \subset M$ be countable and dense, and let $H : M \to [0, 1]^{\mathbb{N}}$ be the map defined by

$$H(x) := (d(x, a_n))_{n \geq 0}.$$

By Tychonoff's theorem (see, e.g., Theorem 2.2.8 in [21]), $[0, 1]^{\mathbb{N}}$ is a compact metric space. A metric for $[0, 1]^{\mathbb{N}}$ is given by

$$e(\mathbf{x}, \mathbf{y}) = \sum_{k \geq 0} \frac{|x_k - y_k|}{2^k},$$

where $\mathbf{x} = (x_k)_{k \geq 0}, \mathbf{y} = (y_k)_{k \geq 0}$. Set

$$\tilde{d}(x, y) := e(H(x), H(y)).$$

It is not hard to check that $\tilde{d}$ is a metric on M inducing the same topology as d. The spaces $(M, \tilde{d})$ and $(H(M), e)$ are thus isometric. Let $K := \overline{H(M)}$. Then K is compact (as a closed subset of a compact space) and thus, there exists a countable and dense family $\{g_n\} \subset C_b(K)$. Let $f \in U_b(M, \tilde{d})$. Since H is an isometry, the map $f \circ H^{-1} : H(M) \to \mathbb{R}$ is uniformly continuous. It then extends to a continuous map $\hat{f} : \overline{H(M)} \to \mathbb{R}$. By density of $\{g_n\}$, there exists, for all $\varepsilon > 0$, some n such that

$$\|f - g_n \circ H\|_\infty = \sup_{\mathbf{x} \in H(M)} |f \circ H^{-1}(\mathbf{x}) - g_n(\mathbf{x})| \leq \sup_{\mathbf{x} \in K} |\hat{f}(\mathbf{x}) - g_n(\mathbf{x})| \leq \varepsilon.$$

This proves that the sequence $\{f_n\}$, with $f_n := g_n \circ H$, is dense in $U_b(M, \tilde{d})$. Now, by Theorem 4.1 (b) and density of $\{f_k\}$, $\mu_n \Rightarrow \mu$ if and only if $\mu_n f_k \to \mu f_k$ for all $k \in \mathbb{N}$. This is equivalent to $D(\mu_n, \mu) \to 0$. $\square$

One of the main advantages of the distance defined in Proposition 4.5 is that it allows to verify weak convergence by testing the condition $\mu_n f \to \mu f$ over a countable set of functions.

Two other classical distances over $\mathcal{P}(M)$ are the following:

Prohorov Metric For any $A \subset M$ and $\varepsilon > 0$, let

$$A^\varepsilon := \{y \in M : d(y, A) < \varepsilon\}.$$

For all $\mu, \nu \in \mathcal{P}(M)$ the *Prohorov* distance (also called the Lévy-Prohorov distance) between μ and ν is defined as

$$\pi(\mu, \nu) := \inf \{\varepsilon > 0 : \mu(A) \leq \nu(A^\varepsilon) + \varepsilon \text{ for all } A \in \mathcal{B}(M)\}. \tag{4.1}$$

Fortet-Mourier Metric Let $L_b(M) \subset C_b(M)$ be the space of bounded Lipschitz maps equipped with the norm

$$\|f\|_{bl} = \|f\|_\infty + Lip(f),$$

where

$$Lip(f) := \sup \left\{ \frac{|f(x) - f(y)|}{d(x, y)} : (x, y) \in M^2, x \neq y \right\}.$$

For all $\mu, \nu \in \mathcal{P}(M)$ the *Fortet-Mourier* distance between μ and ν is defined as

$$\rho(\mu, \nu) := \sup\{|\mu f - \nu f| : f \in L_b(M), \|f\|_{bl} \leq 1\}. \tag{4.2}$$

Theorem 4.7 *The maps π and ρ are distances on $\mathcal{P}(M)$. Let $\{\mu_n\} \subset \mathcal{P}(M)$ and $\mu \in \mathcal{P}(M)$. The following conditions are equivalent:*

(a) $\mu_n \Rightarrow \mu$;
(b) $\rho(\mu_n, \mu) \to 0$;
(c) $\pi(\mu_n, \mu) \to 0$.

Proof We only prove that $(a) \Leftrightarrow (b)$. For more details and the proof of $(b) \Leftrightarrow (c)$, see Dudley [21]. The implication $(b) \Rightarrow (a)$ follows from assertion (c) of Theorem 4.1. Conversely assume (a). We first assume that M is complete. Fix $\varepsilon > 0$. By Ulam's Theorem (or Prohorov's Theorem 4.13 below), one can choose $K \subset M$ compact such that

$$\mu(K) > 1 - \varepsilon. \tag{4.3}$$

Let $K_\varepsilon = \{x \in M : d(x, K) < \varepsilon\}$. By assertion (e) of Theorem 4.1,

$$\mu_n(K_\varepsilon) > 1 - \varepsilon \tag{4.4}$$

for n sufficiently large. By the Arzelà–Ascoli theorem, the unit ball $L_{b,1} := \{f \in L_b : \|f\|_{bl} \leq 1\}$ restricted to K is a compact subset of $C_b(K)$. There exists then a finite set $\{f_1, \ldots, f_N\} \subset L_{b,1}$ such that for all $f \in L_{b,1}$ there is some $i \in \{1, \ldots, N\}$ such that $|f(x) - f_i(x)| \leq \varepsilon$ for all $x \in K$. Since f and f_i have a Lipschitz constant ≤ 1, we also get that

$$|f(x) - f_i(x)| \leq 3\varepsilon \tag{4.5}$$

for all $x \in K_\varepsilon$. Now

$$|\mu_n f - \mu f| \leq |(\mu_n - \mu)f_i| + |(\mu_n - \mu)((f - f_i)\mathbf{1}_{K_\varepsilon})| + |(\mu_n - \mu)((f - f_i)\mathbf{1}_{M \setminus K_\varepsilon})|.$$

Thus, using inequalities (4.3), (4.4), and (4.5), we obtain

$$\rho(\mu_n, \mu) \leq \max_{1 \leq i \leq N} |(\mu_n - \mu)f_i| + 8\varepsilon.$$

This proves (b) for M complete. If M is not complete, we can replace it by its completion $\tilde{M}$. Any map $f \in L_b$ extends to a bounded Lipschitz map on $\tilde{M}$ and the measures (μ_n) and μ can be seen as measures on $\tilde{M}$ so that the proof goes through. $\square$

Remark 4.8 Theorem 4.1 is true in any (not necessarily separable) metric space. The equivalences in Theorem 4.7 require separability (but not completeness).

Exercise 4.9 Let K be a compact metric space (and thus also a Polish space). Using the proof of Proposition 4.5, show that K is homeomorphic to a compact subset of $[0, 1]^{\mathbb{N}}$, equipped with the metric e. We now identify K with a subset of $[0, 1]^{\mathbb{N}}$. Let P be the set of real-valued functions on $[0, 1]^{\mathbb{N}}$ of the form $p(\mathbf{x}) = q(x_0, \ldots, x_n)$, where $q : [0, 1]^{n+1} \to \mathbb{R}$ is a polynomial in $(n + 1)$ variables with rational coefficients. Use the Stone–Weierstrass theorem to show that $P|_K = \{p|_K : p \in P\}$ is dense in $C(K)$. This shows that $C(K)$ is separable. Since $C_b(K)$ is a subset of the separable metric space $C(K)$, it is itself separable.

Exercise 4.10 Let $\mathcal{X}$ be a topological space. Suppose that there exists an uncountable family $\{O_\alpha\}$ of open sets such that $O_\alpha \cap O_\beta = \emptyset$ for $\alpha \neq \beta$. Show that $\mathcal{X}$ is not separable. Show that $C_b(\mathbb{R})$, the set of continuous bounded functions on $\mathbb{R}$, is not separable. *Hint:* Let $f \in C_b(\mathbb{R})$ be such that $f(n) = 0$ and $f = 1$ on $[n + 1/(n + 1), n + 1 - 1/(n + 1)]$ for all $n \in \mathbb{N}^*$. Set $f_x(t) := f(x + t)$ and consider the family $\{O_x\}_{x \in (0,1)}$, where $O_x := \{g \in C_b(\mathbb{R}) : \|f_x - g\|_\infty < 1/2\}$.

Exercise 4.11 (Borel Isomorphism) We say that two measurable spaces X and Y are isomorphic if there exists a bi-measurable bijection $\Psi : X \to Y$, meaning that both Ψ and Ψ^{-1} are measurable. It turns out that every Borel subset M of a Polish space is isomorphic to a Borel subset of $[0, 1]$ (see Remark 4.12). The purpose of this exercise is to prove this result when M is compact or locally compact and separable.

(i) Let $\{0, 1\}^{\mathbb{N}^*}$ be equipped with the product topology and Borel σ-field. Show that $\{0, 1\}^{\mathbb{N}^*}$ is a metric space with the metric d defined as

$$d(\omega, \alpha) := \sum_{i \geq 1} \frac{|\omega_i - \alpha_i|}{2^i}.$$

(ii) Show that the map

$$\Psi : \{0, 1\}^{\mathbb{N}^*} \to [0, 1],$$

$$\omega \mapsto \sum_{i \geq 1} \frac{\omega_i}{2^i}$$

is 1-Lipschitz continuous.

(iii) Let $\tilde{I} \subset \{0, 1\}^{\mathbb{N}^*}$ be the set of ω such that $\omega_i = 0$ for infinitely many i and $\omega_j = 1$ for infinitely many j. Show that $\tilde{I}$ is a Borel subset of $\{0, 1\}^{\mathbb{N}^*}$ and that $\Psi|_{\tilde{I}}$ (Ψ restricted to $\tilde{I}$) is a homeomorphism onto $\Psi(\tilde{I})$, i.e., a continuous bijection with continuous inverse.

(iv) Show that $[0, 1]$ and $\{0, 1\}^{\mathbb{N}^*}$ are isomorphic. *Hint:* Use (iii) and the fact that the complement of $\tilde{I}$ in $\{0, 1\}^{\mathbb{N}^*}$ is countably infinite.

(v) Show that there is a homeomorphism between $\{0, 1\}^{\mathbb{N}^*}$ and $\{0, 1\}^{\mathbb{N}^* \times \mathbb{N}^*}$, equipped with the metric

$$e(A, B) := \sum_{j \geq 1} \frac{d((A_{i,j})_{i \geq 1}, (B_{i,j})_{i \geq 1})}{2^j}.$$

Then show that $[0, 1]$ and $[0, 1]^{\mathbb{N}^*}$ are isomorphic. Relying on the proof of Proposition 4.5, deduce that every compact (or locally compact separable) metric space is isomorphic to a Borel subset of $[0, 1]$. *Hint:* Any locally compact separable metric space can be written as a countable union of compact sets, see, e.g., Theorem XI.6.3 in [23].

Remark 4.12 Theorem 13.1.1 in [21] implies the following: If M is a Borel subset of a Polish space, and if B is a Borel subset of $[0, 1]$ whose cardinality equals the cardinality of M, then M and B are isomorphic. Since the cardinality of a Borel subset of a Polish space is either finite, countably infinite, or the cardinality of the continuum, every such set is in fact isomorphic to a large class of Borel subsets of $[0, 1]$.

4.1.1 Tightness and Prohorov's Theorem

A set $\mathcal{P} \subset \mathcal{P}(M)$ is called *tight* (sometimes uniformly tight) if for every $\varepsilon > 0$ there exists a compact set $K \subset M$ such that

$$\mu(K) \geq 1 - \varepsilon$$

for all $\mu \in \mathcal{P}$. Observe in particular that if M is compact, every subset of $\mathcal{P}(M)$ is tight. A set $\mathcal{P} \subset \mathcal{P}(M)$ is called *relatively compact* if it has compact closure in $\mathcal{P}(M)$ (equipped with one of the distances π, ρ, or any other distance characterizing weak convergence). Finally, it is called *totally bounded* if for every $\varepsilon > 0$ there is a finite set $A \subset \mathcal{P}$ such that the following holds: For every $\mu \in \mathcal{P}$ there is $\nu \in A$ with $d(\mu, \nu) < \varepsilon$. Here, d can be the Prohorov metric, the Fortet-Mourier metric, or any other metric on $\mathcal{P}(M)$ characterizing weak convergence.

The following theorem usually referred to as Prohorov's theorem asserts that tightness and relative compactness are equivalent in a Polish space (complete and separable metric space). Here the assumption that M is a Polish space is crucial, for otherwise the implication $(b) \Rightarrow (a)$ may be false. See, e.g., Billingsley [11] or Dudley [21, Chapter 11.5] for a proof of Prohorov's theorem.

Theorem 4.13 (Prohorov's Theorem) *Assume M is a Polish space (i.e., a complete separable metric space). Then the following assertions are equivalent:*

(a) $\mathcal{P}$ is tight;
(b) $\mathcal{P}$ is relatively compact;
(c) Every sequence $\{\mu_n\} \subset \mathcal{P}$ has a convergent subsequence $\mu_{n_k} \Rightarrow \mu \in \mathcal{P}(M)$;
(d) $\mathcal{P}$ is totally bounded for π or ρ.

Remark 4.14 The latter property shows that $\mathcal{P}(M)$ is complete for ρ or π since every Cauchy sequence is totally bounded.

A Tightness Criterion

We conclude this subsection with a simple practical Lyapunov-type condition ensuring tightness of a sequence of probability measures.

A measurable map $V : M \to \mathbb{R}$ is called *proper* if for all $R \in \mathbb{R}$ the set

$$\{V \le R\} = \{x \in M : V(x) \le R\}$$

has compact closure.

Proposition 4.15 *Let $V : M \to \mathbb{R}^+$ be a proper map and let $\{\mu_n\}$ be a sequence in $\mathcal{P}(M)$ such that*

$$\limsup_{n \to \infty} \mu_n V \le K < \infty.$$

Then $\{\mu_n\}$ is tight. Assume furthermore that V is continuous. Then

(i) For every limit point μ of $\{\mu_n\}$, $\mu V \le K$;
(ii) Let $H : M \to \mathbb{R}$ be a continuous function such that $G = \frac{V}{1+|H|}$ is proper. If $\mu_n \Rightarrow \mu$, then $\mu_n H \to \mu H$.

Proof Fix $\varepsilon > 0$ and let $R > 0$ be so large that $\limsup_{n \to \infty} \mu_n V \le \varepsilon R$. By the Markov inequality, $\limsup_{n \to \infty} \mu_n \{V > R\} \le \limsup_{n \to \infty} \frac{\mu_n V}{R} \le \varepsilon$. Let now $\mu = \lim \mu_{n_k}$ be a limit point of $\{\mu_n\}$. Then for all $R > 0$, $\mu(V \wedge R) = \lim_{k \to \infty} \mu_{n_k}(V \wedge R) \le K$. Thus $\mu V \le K$ by monotone convergence.

We pass to the proof of (ii). Let $G = \frac{V}{1+|H|}$. For all $R \in \mathbb{R} \setminus D$ with D at most countable, $\mu\{G = R\} = 0$ and, therefore,

$$\lim_{n \to \infty} \mu_n (H \mathbf{1}_{G \le R}) = \mu(H \mathbf{1}_{G \le R}).$$

On the other hand $\mu_n(|H| \mathbf{1}_{G > R}) \le \mu_n(\frac{V}{G} \mathbf{1}_{G > R}) \le \frac{1}{R} \mu_n(V)$. Thus

$$\lim_{R \to \infty} \limsup_{n \to \infty} \mu_n(|H| \mathbf{1}_{G > R}) = 0$$

and, similarly,

$$\lim_{R \to \infty} \mu(|H| \mathbf{1}_{G > R}) = 0.$$

This proves the result.   □

4.2 Invariant Measures

Given a Markov kernel P, a measure (respectively a probability measure) μ is called *P-invariant* or simply *invariant* if

$$\mu P f = \mu f \qquad (4.6)$$

for all $f \in B(M)$, where Pf is defined by (1.1). Equivalently,

$$\mu P = \mu,$$

where μP is defined by (1.2).

Exercise 4.16 Let $\mathbb{R}/\mathbb{Z}$ denote the set of equivalence classes with respect to the equivalence relation $x \sim y \Leftrightarrow x - y \in \mathbb{Z}$ on $\mathbb{R}$. The set $\mathbb{R}/\mathbb{Z}$ can be thought of as the unit interval $[0, 1]$, where 0 and 1 are identified with each other. Let $(\theta_n)_{n \geq 1}$ be an i.i.d. sequence of random variables with distribution m, where m is a Borel probability measure on $\mathbb{R}/\mathbb{Z}$. For every $\theta \in \mathbb{R}$, let

$$F_\theta : \mathbb{R}/\mathbb{Z} \to \mathbb{R}/\mathbb{Z}, \ x \mapsto x + \theta \mod 1.$$

Show that the Lebesgue measure on $\mathbb{R}/\mathbb{Z}$ is an invariant probability measure for the Markov kernel induced by (F, m).

Remark 4.17 Let $\mathcal{C}$ denote a set of bounded, measurable mappings $f : M \to \mathbb{R}$, closed under multiplication and such that $\mathcal{B}(M) = \sigma(\mathcal{C})$ (the smallest σ-field making elements of $\mathcal{C}$ measurable). By a monotone class argument (see Theorem A.1), it suffices to check (4.6) on $\mathcal{C}$ to prove P-invariance of $\mu \in \mathcal{P}(M)$.

For instance, one can choose $\mathcal{C} = C_b(M)$, the set of bounded continuous functions. One can also choose any set $\mathcal{C} \subset C_b(M)$ closed under multiplication and such that for all $f \in C_b(M)$ there is a sequence $\{f_n\} \subset \mathcal{C}$ such that $\lim_{n \to \infty} f_n(x) = f(x)$ for all $x \in M$.

We let $\mathsf{Inv}(P)$ denote the set of P-invariant probability measures. The set $\mathsf{Inv}(P)$ might be empty as shown by the following two examples.

Example 4.18 Let $M = [0, 1]$ and $f : M \to M$ be the map defined by $f(x) = x/2$ for $x \neq 0$ and $f(0) = 1$. Then the (deterministic) chain $X_{n+1} = f(X_n)$ has no invariant probability measure. For otherwise the Poincaré recurrence theorem (see Theorem 4.41 below) would imply that such a measure is δ_0, but $f(0) = 1$.

Example 4.19 Consider the pair (F, m) introduced in Exercise 4.16. Let us assume in addition that $\int_{\mathbb{R}} |\theta| \ m(d\theta) < \infty$ and set $\alpha := \int_{\mathbb{R}} \theta \ m(d\theta)$. While the corresponding Markov kernel P has the Lebesgue measure as an invariant measure, P does not admit any invariant probability measures if $\alpha \neq 0$.

To see this, let μ be a probability measure on $(\mathbb{R}, \mathcal{B}(\mathbb{R}))$. Then there is $K > 0$ such that $\mu([-K, K]) > 0$. If μ was invariant for P, the Markov chain $(X_n)_{n \in \mathbb{N}}$

induced by (F, m) and with $X_0 \sim \mu$ would satisfy

$$0 < \mu([-K, K]) = \mu P^n([-K, K]) = \mathsf{P}(|X_n| \leq K), \quad \forall n \in \mathbb{N}^*.$$

But if $\alpha > 0$ ($\alpha < 0$), one has $\lim_{n\to\infty} X_n = \infty$ ($\lim_{n\to\infty} X_n = -\infty$) P-almost surely by the law of large numbers. Hence $\lim_{n\to\infty} \mathsf{P}(|X_n| \leq K) = 0$, a contradiction.

Given a Markov chain (X_n) on M, the associated family of *empirical occupation measures* is defined as

$$\nu_n := \frac{1}{n} \sum_{i=0}^{n-1} \delta_{X_i}, \quad n \in \mathbb{N}^*. \tag{4.7}$$

Notice that each ν_n is a random element of $\mathcal{P}(M)$.

A sufficient condition ensuring existence of invariant probability measures is given by the following classical theorem (see, e.g., [22]).

Theorem 4.20 *Let (X_n) denote a Feller Markov chain (defined on $(\Omega, \mathcal{F}, \mathsf{P})$) on M with kernel P. Then the following statements hold.*

(i) P-*almost surely, every limit point of the family of empirical occupation measures $(\nu_n)_{n \geq 1}$ is P-invariant;*
(ii) *If $(\nu_n)_{n \geq 1}$ is tight with positive P-probability, then $\mathsf{Inv}(P)$ is nonempty.*

Proof

(i) Let $f \in B(M)$. Set $U_{n+1} := f(X_{n+1}) - Pf(X_n)$, $M_0 := 0$, and $M_{n+1} := M_n + U_{n+1}$ for $n \geq 0$. Then (M_n) is an L^2-martingale, whose predictable quadratic variation (see the section on martingale theory in the appendix) verifies

$$\langle M \rangle_{n+1} - \langle M \rangle_n = \mathsf{E}(U_{n+1}^2 | \mathcal{F}_n) = Pf^2(X_n) - (Pf)^2(X_n) \leq 2\|f\|_\infty^2.$$

Hence by the strong law of large numbers for martingales (see Theorem A.8),

$$0 = \lim_{n\to\infty} \frac{M_n}{n} = \lim_{n\to\infty} \nu_n f - \nu_n(Pf) \tag{4.8}$$

almost surely. Let $\{f_k\} \subset C_b(M)$ be as in Proposition 4.5. Then, by the Feller property, Pf_k is in $C_b(M)$ for all k and, consequently, with probability one

$$\nu f_k - \nu(Pf_k) = 0$$

for every limit point ν of $\{\nu_n\}$ and every $k \in \mathbb{N}$. Thus, $\nu = \nu P$.

(ii) Let $\omega \in \Omega$ such that $(\nu_n(\omega))_{n \geq 1}$ is tight and all of its limit points are P-invariant. By Prohorov's theorem, $(\nu_n(\omega))_{n \geq 1}$ admits at least one limit point, so $\mathsf{Inv}(P)$ is nonempty.

$\square$

Corollary 4.21 *If M is compact and P is Feller, $\mathsf{Inv}(P)$ is a nonempty compact convex subset of $\mathcal{P}(M)$. Convexity of $\mathsf{Inv}(P)$ holds for arbitrary metric spaces and Markov kernels.*

4.2.1 Tightness Criteria for Empirical Occupation Measures

When M is noncompact, the tightness of the empirical occupation measures (ν_n) can be ensured by the existence of a convenient Lyapunov function. This is a proper map $V : M \to \mathbb{R}_+$ such that $PV - V$ is "sufficiently" negative.

Corollary 4.22 *Let $V : M \to \mathbb{R}_+$ be a proper map. Assume that $PV \leq V$ and that $\mathsf{E}(V(X_0)) < \infty$. Then the family of empirical occupation measures (ν_n) is almost surely tight.*

Proof The sequence $\{V_n = V(X_n)\}$ being a nonnegative supermartingale with $\mathsf{E}(V_0) < \infty$, it converges almost surely to some finite random variable V_∞ (see Theorem A.6). This implies that $\nu_n V \to V_\infty$ almost surely and the result follows from Proposition 4.15. $\square$

Another result, in the same spirit, is

Corollary 4.23 *Let $V : M \to \mathbb{R}_+$ be a proper map. Assume that*

$$PV \leq \rho V + \kappa,$$

with $\kappa \geq 0, 0 \leq \rho < 1$, and $\mathsf{E}(V(X_0)) < \infty$. Then

$$\limsup_{n \to \infty} \nu_n \sqrt{V} \leq \frac{\sqrt{\kappa}}{1 - \sqrt{\rho}}$$

almost surely. In particular, (ν_n) is tight. The set $\mathsf{Inv}(P)$ is a nonempty compact convex subset of $\mathcal{P}(M)$ and for all $\mu \in \mathsf{Inv}(P)$, $\mu V \leq \frac{\kappa}{1-\rho}$.

Proof Set $W = \sqrt{V}$. Then, by Jensen's inequality,

$$PW \leq \sqrt{PV} \leq \sqrt{\rho V + \kappa} \leq \sqrt{\rho}W + \sqrt{\kappa}.$$

Set $LW(x) = PW(x) - W(x)$, $M_0 = 0$, and

$$M_n = W(X_n) - W(X_0) - \sum_{k=0}^{n-1} LW(X_k)$$

for all $n \geq 1$. Then (M_n) is an L^2-martingale whose predictable quadratic variation process is given as $\langle M \rangle_0 = 0$ and

$$\langle M \rangle_{n+1} - \langle M \rangle_n = \mathsf{E}((M_{n+1} - M_n)^2 | \mathcal{F}_n) = PV(X_n) - (PW)^2(X_n) \leq PV(X_n)$$

for $n \geq 0$. Thus $\mathsf{E}(\langle M \rangle_n) \leq \sum_{i=0}^n \mathsf{E}(P^{i+1}V(X_0)) \leq n\frac{\kappa}{1-\rho} + \frac{\rho}{1-\rho}\mathsf{E}(V(X_0))$, where the last inequality easily follows from the assumptions on V. Then, by the second strong law of large numbers for L^2-martingales (Theorem A.8 (iv)), $\frac{M_n}{n} \to 0$ almost surely. Now, because $-LW \geq (1 - \sqrt{\rho})W - \sqrt{\kappa}$,

$$(1 - \sqrt{\rho})v_n W \leq \sqrt{\kappa} + \frac{M_n}{n} + \frac{W(X_0)}{n}.$$

This, combined with Proposition 4.15, proves the first statement.

By Theorem 4.20, $\mathsf{Inv}(P)$ is nonempty. Let $\mu \in \mathsf{Inv}(P)$. For all $n \in \mathbb{N}^*$,

$$P^n V \leq \rho^n V + \kappa\frac{1 - \rho^n}{1 - \rho} \leq \rho^n V + \kappa\frac{1}{1 - \rho}.$$

Thus, by invariance and Jensen's inequality,

$$\mu(V \wedge M) = \mu P^n(V \wedge M) \leq \mu(P^n V \wedge M) \leq \mu((\rho^n V + \frac{\kappa}{1 - \rho}) \wedge M).$$

Letting $n \to \infty$ in the right-hand term and using dominated convergence shows that $\mu(V \wedge M) \leq \frac{\kappa}{1-\rho}$. Then $\mu V \leq \frac{\kappa}{1-\rho}$ by monotone convergence. Compactness follows from Proposition 4.15 and Prohorov's theorem. □

Exercise 4.24 (Invariant Measures and Mean-Occupation)   Let (X_k) be a Markov chain, T a finite stopping time (i.e., $T < \infty$ a.s.) and let ν be the "mean occupation measure up to time T" defined for all $f \in B(M)$, $f \geq 0$, as

$$\nu f := \mathsf{E}\left(\sum_{k=0}^{T-1} f(X_k) \right).$$

(i) Show that $\nu(Pf) - \nu f = \mathsf{E}(f(X_T)) - \mathsf{E}(f(X_0))$.
(ii) Show that if X_0 and X_T have the same distribution and $\mathsf{E}(T) < \infty$, then $\frac{\nu}{\nu(1)}$ is an invariant probability measure for the chain.

4.3 Excessive Measures

A measure μ is called *excessive* provided

$$\mu P \leq \mu.$$

Lemma 4.25 *Every finite excessive measure is invariant.*

Proof If μ is a finite excessive measure, then $\mu P(A) \leq \mu(A)$ and $\mu(M) - \mu P(A) = \mu P(A^c) \leq \mu(A^c) = \mu(M) - \mu(A)$, so that $\mu P(A) = \mu(A)$. □

Given two Borel measures α and β on M, one calls α *absolutely continuous* with respect to β and writes $\alpha \ll \beta$ if for every $A \in \mathcal{B}(M)$, $\beta(A) = 0$ implies that $\alpha(A) = 0$. One says that α and β are *mutually singular* and writes $\alpha \perp \beta$ if there is $A \in \mathcal{B}(M)$ such that $\alpha(A) = \beta(A^c) = 0$. Let μ and ν be Borel measures on M. By Lebesgue's decomposition theorem (see, e.g., Theorem 3.8 in [27]), $\nu = \nu_{ac} + \nu_s$, where $\nu_{ac} \ll \mu$ and $\nu_s \perp \mu$. Equivalently,

$$\nu(dx) = h(x)\mu(dx) + \mathbf{1}_A(x)\nu(dx),$$

where $h \in L^1(\mu)$ and $\mu(A) = 0$.

Lemma 4.26 *Let $\mu, \nu \in \mathsf{Inv}(P)$. Then the absolutely continuous and the singular parts of ν with respect to μ are invariant measures.*

Proof Write $\nu(dx) = h(x)\mu(dx) + \mathbf{1}_A(x)\nu(dx)$ with $h \in L^1(\mu)$ and $\mu(A) = 0$. By invariance, $\mu(A) = \int P(x, A)\mu(dx) = 0$, so that $P(x, A) = 0$ for μ-almost every $x \in M$. Thus, for every Borel set B,

$$\int P(x, B)h(x)\,\mu(dx) = \int P(x, B \cap A^c)h(x)\,\mu(dx) \leq \nu(B \cap A^c) = (h\mu)(B).$$

This proves that $h(x)\mu(dx)$ is finite and excessive, hence invariant. Since $\mathbf{1}_A(x)\nu(dx) = \nu(dx) - h(x)\mu(dx)$ and since $\nu \in \mathsf{Inv}(P)$, the measure $\mathbf{1}_A(x)\nu(dx)$ is invariant as well. □

4.4 Ergodic Measures

Let $\mu \in \mathsf{Inv}(P)$. A bounded, measurable function g is called (P, μ)-invariant provided $Pg = g$, μ-almost surely. A set $A \in \mathcal{B}(M)$ is called (P, μ)-invariant if $\mathbf{1}_A$ is (P, μ)-invariant.

An invariant probability measure μ is called *ergodic* (for P) if every (P, μ)-invariant function is μ-almost surely constant. (A function $f : M \to \mathbb{R}$ is called

μ-almost surely constant if there is $c \in \mathbb{R}$ such that $f(x) = c$ for μ-almost every $x \in M$.)

Lemma 4.27 *A probability measure* $\mu \in \mathsf{Inv}(P)$ *is ergodic if and only if every* (P, μ)-*invariant set has* μ-*measure* 0 *or* 1.

Proof Suppose first that $\mu \in \mathsf{Inv}(P)$ is not ergodic. Then there exists a bounded, measurable function h such that $Ph = h$, μ-almost surely, and for every $c \in \mathbb{R}$

$$\mu(\{x \in M : h(x) = c\}) < 1.$$

It follows that for some $c \in \mathbb{R}$, $A := \{x \in M : h(x) > c\}$ has μ-measure different from 0 and 1.

Claim A is (P, μ)-invariant.

Proof of the Claim By Jensen's inequality, $|Ph| \le P|h|$. Since $\mu(P|h| - |h|) = 0$ by P-invariance of μ, and since $Ph = h$ μ-almost surely, this proves that $|h|$ is (P, μ)-invariant as well. Hence, $\max(0, h) = \frac{1}{2}(h + |h|)$ is (P, μ)-invariant. Similarly,

$$h_n := \min(n \max(0, h - c), 1)$$

is (P, μ)-invariant for every $n \ge 1$. Since $\lim_{n \to \infty} h_n = \mathbf{1}_A$, $\mathbf{1}_A$ is (P, μ)-invariant as the pointwise limit of a uniformly bounded sequence of (P, μ)-invariant functions. This proves the claim and one direction of the lemma.

For the converse direction, let μ be ergodic and let A be a (P, μ)-invariant set. Then $\mathbf{1}_A$ is a (P, μ)-invariant function, and ergodicity of μ implies that there is $c \in \mathbb{R}$ such that $\mathbf{1}_A$ is μ-almost surely equal to c. Necessarily, $c \in \{0, 1\}$, whence it follows that $\mu(A) \in \{0, 1\}$.

$\square$

Remark 4.28 One usually defines a *harmonic* map as a measurable map (bounded or nonnegative) such that $Pf = f$. Note that a harmonic map is (P, μ)-invariant for every $\mu \in \mathsf{Inv}(P)$.

A probability measure $\mu \in \mathsf{Inv}(P)$ is called *extremal* if it cannot be written as $\mu = (1 - t)\mu_0 + t\mu_1$ with $\mu_0, \mu_1 \in \mathsf{Inv}(P), 0 < t < 1$, and $\mu_0 \ne \mu_1$. Notice that an extremal invariant probability measure cannot be written as the sum of two nontrivial invariant measures that are mutually singular. This fact will be used below in the proof of Proposition 4.29 (ii).

Proposition 4.29

 (i) *An invariant probability measure* μ *is ergodic if and only if it is extremal.*
(ii) *Two distinct ergodic measures are mutually singular.*

Proof

(i) Suppose that μ is nonergodic. By Lemma 4.27, there exists a (P, μ)-invariant set A such that $0 < \mu(A) < 1$. Let $\mu_A(\cdot) := \mu(A \cap \cdot)/\mu(A)$. We claim that for every $g \in B(M)$,

$$P(g\mathbf{1}_A) = (Pg)\mathbf{1}_A$$

μ-almost surely. Indeed, by the Cauchy–Schwarz inequality,

$$|P(g\mathbf{1}_A)|^2 \le P(g^2)P(\mathbf{1}_A) = P(g^2)\mathbf{1}_A$$

μ-almost surely. Thus $P(g\mathbf{1}_A)\mathbf{1}_{A^c} = 0$, μ-almost surely, and, interchanging the roles of A and A^c, $P(g\mathbf{1}_{A^c})\mathbf{1}_A = 0$, μ-almost surely. On the other hand,

$$P(g\mathbf{1}_A) - (Pg)\mathbf{1}_A = [P(g\mathbf{1}_A) - Pg]\mathbf{1}_A + P(g\mathbf{1}_A)\mathbf{1}_{A^c}$$

$$= -P(g\mathbf{1}_{A^c})\mathbf{1}_A + P(g\mathbf{1}_A)\mathbf{1}_{A^c} = 0.$$

This proves the claim. Therefore,

$$\mu_A(Pg) = \frac{1}{\mu(A)}\mu((Pg)\mathbf{1}_A) = \frac{1}{\mu(A)}\mu(P(g\mathbf{1}_A))$$

$$= \frac{1}{\mu(A)}\mu(g\mathbf{1}_A) = \mu_A(g).$$

This proves that μ_A is an invariant probability measure. Similarly, μ_{A^c} is an invariant probability measure, and since $\mu = \mu(A)\mu_A + (1 - \mu(A))\mu_{A^c}$, the probability measure μ is nonextremal.

Suppose now that μ is ergodic and that $\mu = (1-t)\mu_0 + t\mu_1$ with $\mu_0, \mu_1 \in \mathsf{Inv}(P)$ and $t \in [0, 1]$. If $t \ne 0$, $\mu_1 \ll \mu$. Hence, there exists $h \in L^1(\mu)$ such that $\mu_1(dx) = h(x)\mu(dx)$. Furthermore, $h \le 1/t$, μ-almost surely, because for all $c > 0$,

$$tc\mu\{h \ge c\} \le t\mu_1\{h \ge c\} \le \mu\{h \ge c\}.$$

In particular, h and Ph lie in $L^2(\mu)$. Then, by Jensen's inequality,

$$0 \le \mu((Ph - h)^2) = \mu((Ph)^2 - 2hPh + h^2) \le \mu(Ph^2 - 2hPh + h^2)$$

$$= 2\mu h^2 - 2\mu_1 Ph = 2\mu h^2 - 2\mu_1 h = 0,$$

from which it follows that $Ph = h$, μ-almost surely, and, by ergodicity and the fact that μ and μ_1 are probability measures, $h = 1$. As a result, $t = 1$ and μ is extremal.

(ii) Let μ and v be ergodic. Write $v(dx) = h(x)\mu(dx) + \mu_s(dx)$ with μ_s singular with respect to μ and $h \in L^1(\mu)$. By Lemma 4.26, $h(x)\mu(dx)$ and μ_s are invariant, and by extremality either $h = 0$ or $\mu_s = 0$. If $h = 0$, we are done. If $\mu_s = 0$, we claim that $Ph = h$, μ-almost surely. Thus, by ergodicity, $h = 1$, μ-almost surely. This yields $\mu = v$ and we are done. The proof of the claim is easy if $h \in L^2(\mu)$ because, reasoning exactly as in the end of the proof of (i), one finds that $\mu(Ph - h)^2 = 0$. If now $h \in L^1(\mu) \setminus L^2(\mu)$, set $h_n = h \wedge n$ and $\mu_n = h_n\mu$. Then, for all $A \in \mathcal{B}(M)$, $\mu_n P(A) \le n\mu P(A) = n\mu(A)$ and $\mu_n(A) \le vP(A) = v(A)$. Thus

$$\mu_n P(A) = \mu_n P(A \cap \{h \le n\}) + \mu_n P(A \cap \{h > n\})$$

$$\le v(A \cap \{h \le n\}) + n\mu(A \cap \{h > n\}) = \mu_n(A).$$

This shows that μ_n is excessive, hence invariant, by Lemma 4.25. Thus, $Ph_n = h_n$, by what precedes, and $Ph = h$, μ-almost surely, by monotone convergence.

$\square$

4.5 Unique Ergodicity

We say that (X_n) or P is *uniquely ergodic* if the set of P-invariant probability measures has cardinality one. An immediate consequence of the preceding section is

Proposition 4.30 *If P is uniquely ergodic, then its invariant probability measure is ergodic.*

While a deterministic dynamical system is rarely uniquely ergodic (see Sect. 4.6 for a definition of ergodic probability measures for deterministic dynamical systems), this property is much more often satisfied by random dynamical systems and Markov chains. We start with a simple situation, which can be seen as a random version of the Banach fixed point theorem.

4.5.1 Unique Ergodicity of Random Contractions

Throughout this subsection, let M be a complete, separable metric space. Recall that a map $f : M \to M$ is a *contraction* if its Lipschitz constant

$$Lip(f) := \sup \left\{ \frac{d(f(x), f(y))}{d(x, y)} : x \neq y \right\}$$

is < 1. By the Banach fixed point theorem, a contraction f has a unique fixed point x^*, and for all $x \in M$, $f^n(x) \to x^*$ at an exponential rate. Here, using the notation of Chap. 3, we shall consider a Markov chain recursively defined by

$$X_{n+1} = F_{\theta_{n+1}}(X_n)$$

under the assumption that the maps F_θ are contracting on average.

More precisely, we assume that for each $\theta \in \Theta$, the map F_θ is Lipschitz continuous, and we let $l_\theta := Lip(F_\theta)$. Note that, by separability, the supremum in the definition of the Lipschitz constant can be chosen over a countable set, so that l_θ is measurable in θ.

We say that the family $\{F_\theta\}$ is *contracting on average* if $\int \log(l_\theta)^+ m(d\theta) < \infty$ and

$$\int \log(l_\theta) \, m(d\theta) =: -\alpha < 0.$$

Here, we allow for α to be $+\infty$. The next result is classical and has been proved in several places. Here we follow the approach of Diaconis and Freedman [19].

Theorem 4.31 *Assume that $\{F_\theta\}$ is contracting on average and that*

$$\int \log(d(F_\theta(x_0), x_0))^+ m(d\theta) < \infty \tag{4.9}$$

for some $x_0 \in M$. Then the induced Markov chain has a unique invariant probability measure μ^*, and X_n converges in distribution to μ^*. In other words, for every probability measure μ on M,

$$\mu P^n \Rightarrow \mu^*.$$

If we furthermore assume that $\alpha < \infty$,

$$A := \sup_\theta | \log(l_\theta) + \alpha | < \infty$$

and

$$B := \int d(F_\theta(x_0), x_0)\, m(d\theta) < \infty,$$

then for every $x \in M$ there is $C(x) > 0$ such that

$$\rho(\delta_x P^n, \mu^*) \leq C(x)e^{-n\beta}, \quad \forall n \in \mathbb{N},$$

where ρ stands for the Fortet-Mourier distance (see (4.2)),
and $\beta := \min\{\alpha/4, \alpha^2/(32A^2)\}$.

Proof For all $x \in M$, set $X_n^x := F_{\theta_n} \circ \ldots \circ F_{\theta_1}(x)$ and $Y_n^x := F_{\theta_1} \circ \ldots \circ F_{\theta_n}(x)$.
The idea of the proof is to show that (Y_n^x) converges almost surely (and thus in law)
to some random variable Y_∞ independent of x. Since X_n^x and Y_n^x have the same law,
this implies that (X_n^x) converges in law to Y_∞.

To shorten notation, set $l_n := l_{\theta_n}, L_n := \prod_{i=1}^n l_i$, and $Y_n := Y_n^{x_0}$ for x_0 as
in (4.9). By the strong law of large numbers, P-almost surely,

$$\lim_{n\to\infty} \frac{\log(L_n)}{n} = -\alpha \in [-\infty, 0). \tag{4.10}$$

Thus, P-almost surely,

$$\limsup_{n\to\infty} \frac{\log(d(Y_n^x, Y_n))}{n} \leq -\alpha$$

because $d(Y_n^x, Y_n^y) \leq L_n d(x, y)$. We shall now show that (Y_n) is almost surely
Cauchy, by completeness of M hence convergent.

For all $n, p \in \mathbb{N}$,

$$d(Y_{n+p}, Y_n) \leq \sum_{i=0}^{p-1} d(Y_{n+i+1}, Y_{n+i}) \leq \sum_{i\geq 0} L_{n+i} d(F_{\theta_{n+i+1}}(x_0), x_0). \tag{4.11}$$

Let $0 < \varepsilon < \alpha/2$. Then

$$\sum_{n\geq 1} P(\log d(F_{\theta_n}(x_0), x_0) \geq \varepsilon n) \leq \sum_{n\geq 1} P(\log(d(F_{\theta_1}(x_0), x_0))^+ \geq \varepsilon n)$$

$$\leq \frac{1}{\varepsilon} E(\log(d(F_{\theta_1}(x_0), x_0))^+) < \infty.$$

(Here, we used that $\sum_{n\geq 1} P(\xi \geq n) \leq E(\xi)$ for every nonnegative random variable ξ, as well as the integrability condition in (4.9).) Thus, by Borel–Cantelli,

$$\limsup_{n\to\infty} \frac{\log d(F_{\theta_n}(x_0), x_0)}{n} \leq \varepsilon$$

almost surely. Combined with (4.11) and (4.10), it follows that, almost surely, for n large enough,

$$d(Y_{n+p}, Y_n) \leq \sum_{i\geq 0} e^{-(n+i)(\alpha-2\varepsilon)}.$$

This concludes the proof of the first statement, with μ^* the law of the limiting random variable Y_∞ (see also Exercise 4.32).

We now pass to the second statement. For every bounded Lipschitz function f with $\|f\|_{bl} \leq 1$ and for every $\delta > 0$,

$$|\delta_x P^n f - \mu^* f| = |E(f(Y_n^x) - f(Y_\infty))| \leq \delta + 2P(d(Y_n^x, Y_\infty) \geq \delta). \qquad (4.12)$$

First observe that by (4.11),

$$d(Y_n^x, Y_\infty) \leq d(Y_n^x, Y_n) + d(Y_n, Y_\infty) \leq L_n d(x, x_0) + \sum_{i\geq 0} L_{n+i} d(x_0, F_{\theta_{n+i+1}}(x_0)).$$

By Markov's inequality,

$$P(d(x_0, F_{\theta_n}(x_0)) \geq e^{\varepsilon n}) \leq B e^{-\varepsilon n}$$

and by a standard Chernoff inequality (see Exercise 4.33 below),

$$P(L_n \geq e^{(-\alpha+\varepsilon)n}) \leq e^{-n(\varepsilon^2/2A^2)}.$$

Thus

$$P(d(Y_n^x, Y_\infty) \geq e^{(-\alpha+\varepsilon)n} d(x, x_0) + \sum_{i\geq 0} e^{(-\alpha+\varepsilon)(n+i)} e^{\varepsilon(n+i)})$$

$$\leq e^{-n(\varepsilon^2/2A^2)} + \sum_{i\geq 0} \left(e^{-(n+i)(\varepsilon^2/2A^2)} + B e^{-\varepsilon(n+i)} \right).$$

Choose $\varepsilon = \alpha/4$. Then

$$P(d(Y_n^x, Y_\infty) \geq e^{-n\alpha/2}(d(x, x_0) + \frac{1}{1 - e^{-\alpha/2}}))$$

$$\leq e^{-n\alpha^2/(32A^2)}\left(1 + \frac{1}{1 - e^{-\alpha^2/(32A^2)}}\right) + Be^{-n\alpha/4}\frac{1}{1 - e^{-\alpha/4}},$$

and we obtain the desired estimate with the help of (4.12).   □

Exercise 4.32 Let P be a Markov kernel on a separable metric space M, and let μ^* be a Borel probability measure on M such that for every $x \in M$, $\delta_x P^n$ converges weakly to μ^* as $n \to \infty$. Show that if P is Feller, then μ^* is the unique invariant probability measure for P.

Exercise 4.33 (Chernoff Bounds) Let X be an L^1-random variable with zero mean. Assume that $E(e^{\lambda_0 X}) < \infty$ for some $\lambda_0 > 0$. Let $g(\lambda) := \ln(E(e^{\lambda X}))$.

(i) Show that for all $\varepsilon > 0$ and $0 \leq \lambda \leq \lambda_0$,

$$P(X \geq \varepsilon) \leq e^{-\lambda\varepsilon + g(\lambda)}$$

and

$$P(X \geq \varepsilon) \leq e^{-g^*(\varepsilon)},$$

where

$$g^*(\varepsilon) := \sup_{0 \leq \lambda \leq \lambda_0} (\lambda\varepsilon - g(\lambda)).$$

(ii) Assume $|X| \leq A < \infty$. Show that $g(\lambda) \leq \frac{A^2\lambda^2}{2}$ and $g^*(\varepsilon) \geq \frac{\varepsilon^2}{2A^2}$. *Hint:* For the first inequality, it may help to use convexity of g.

(iii) Let (X_n) be a sequence of i.i.d. random variables with the same distribution as X. Show that

$$P(X_1 + \ldots + X_n \geq n\varepsilon) \leq e^{-ng^*(\varepsilon)}.$$

4.6 Classical Results from Ergodic Theory

We first recall some basic definitions from ergodic theory. There are numerous textbooks on the subject including Cornfeld, Fomin, Sinai [17], Mañé [48], Katok and Hasselblatt [42].

Let $(X, \mathcal{F})$ be a measurable space and $T : X \to X$ a measurable mapping. A probability measure $\mathbb{P}$ over X is called T-*invariant* (or simply *invariant*) if

$$\mathbb{P}(T^{-1}(A)) = \mathbb{P}(A)$$

for all $A \in \mathcal{F}$. Given such a $\mathbb{P}$, a measurable function $g : X \to \mathbb{R}$ is called $(T, \mathbb{P})$-*invariant* if $g \circ T = g$, $\mathbb{P}$-almost surely, and a measurable set $A \in \mathcal{F}$ is called $(T, \mathbb{P})$-*invariant* if $\mathbf{1}_A$ is $(T, \mathbb{P})$-invariant. One also defines a T-*invariant set* (or simply *invariant set*) as a set $A \in \mathcal{F}$ such that $T^{-1}(A) = A$. Note that this definition of invariance makes no reference to the measure $\mathbb{P}$ and that a T-invariant set is clearly $(T, \mathbb{P})$-invariant.

A T-invariant probability measure $\mathbb{P}$ is called T-*ergodic* (or simply *ergodic*) provided that every $(T, \mathbb{P})$-invariant function is $\mathbb{P}$-almost surely constant.

Example 4.34 A *periodic point* of period $d \geq 1$ for T is a point $x \in X$ such that $T^d(x) = x$ and $T^i(x) \neq x$ for $i = 1, \ldots, d - 1$. Given such a point, the measure

$$\frac{1}{d}(\delta_x + \delta_{T(x)} + \ldots + \delta_{T^{d-1}(x)})$$

is T-ergodic.

Remark 4.35 One sometimes says that T is *ergodic with respect to* $\mathbb{P}$ to mean that $\mathbb{P}$ is T-ergodic.

Proposition 4.36 *The following assertions are equivalent:*

(a) *The probability measure $\mathbb{P}$ is T-ergodic;*
(b) *Every $(T, \mathbb{P})$-invariant set has $\mathbb{P}$-measure 0 or 1;*
(c) *Every T-invariant set has $\mathbb{P}$-measure 0 or 1.*

Proof The implications $(a) \Rightarrow (b) \Rightarrow (c)$ are obvious. To show that $(c) \Rightarrow (b)$, let A be a $(T, \mathbb{P})$-invariant set. The set

$$\tilde{A} := \{x \in X : T^k(x) \in A \text{ for infinitely many } k \in \mathbb{N}\}$$

is invariant. Hence, by (c), $\mathbb{P}(\tilde{A}) \in \{0, 1\}$. If $x \in A \setminus \tilde{A}$, there exists $k \geq 1$ such that $x \in A \setminus T^{-k}(A)$, and if $x \in \tilde{A} \setminus A$, there exists $k \geq 1$ such that $x \in T^{-k}(A) \setminus A$. It then follows that

$$A \Delta \tilde{A} \subset \bigcup_{k \geq 1} A \Delta T^{-k}(A).$$

Thus

$$\mathbb{P}(A \Delta \tilde{A}) \le \sum_{k \ge 1} \mathbb{P}(A \Delta T^{-k}(A)).$$

Now

$$\mathbb{P}(A \Delta T^{-k}(A)) \le \sum_{i=0}^{k-1} \mathbb{P}(T^{-i}(A) \Delta T^{-(i+1)}(A)) = k\mathbb{P}(A \Delta T^{-1}(A)) = 0.$$

It remains to prove that $(b) \Rightarrow (a)$. Let h be $(T, \mathbb{P})$-invariant. Then, for each $c \in \mathbb{R}$, the set $\{x \in X : h(x) > c\}$ is $(T, \mathbb{P})$-invariant and the result follows.   □

Exercise 4.37 (Rotations) Let $S^1 = \mathbb{R}/\mathbb{Z}$, $\alpha \in S^1$, and $T_\alpha : S^1 \to S^1$ the rotation $x \mapsto x + \alpha$. Describe the invariant and ergodic probability measures of T_α. Show that when α is *irrational* (i.e., $\alpha = \xi + \mathbb{Z}$ with $\xi \in (0, 1) \setminus \mathbb{Q}$), T_α is uniquely ergodic and, more precisely, the normalized Lebesgue measure λ on S^1 is the unique invariant probability measure for T_α.

Exercise 4.38 Let $k \ge 2$ be an integer and $Z^k : S^1 \to S^1$, $x \mapsto kx$. Show that the normalized Lebesgue measure λ is ergodic for Z^k. Show that Z^k has infinitely many periodic points, hence infinitely many ergodic measures.

Exercise 4.39 (Shift) Let $M = \{0, 1\}^{\mathbb{N}^*}$ and let Θ be the shift map on M defined by $\Theta(\omega)_i = \omega_{i+1}$. Show the following statements.

(a) For all $n \ge 1$, Θ has 2^n periodic orbits of period n, and the set of periodic points is dense in M;
(b) There is a point $x \in M$ whose orbit is dense in M;
(c) The probability measure $(\frac{1}{2}(\delta_0 + \delta_1))^{\otimes \mathbb{N}^*}$ is ergodic for Θ;
(d) There exists a continuous surjective map $\Psi : M \to S^1$ such that

$$\Psi \circ \Theta = Z^2 \circ \Psi,$$

where Z^2 is defined as in Exercise 4.38. *Hint:* One can use Exercise 4.11.

Using (d), prove that Z^2 possesses a dense orbit and give an alternative proof of the results of Exercise 4.38 when $k = 2$.

Exercise 4.40 Let $T : S^1 \times S^1 \to S^1 \times S^1$, $(x, y) \mapsto (x+\alpha, y+x)$ with α irrational. Show that $\lambda \otimes \lambda$ is ergodic. *Hint:* One can use the fact that every $f \in L^2(\lambda \otimes \lambda)$ can be written as a Fourier series $f(x, y) = \sum_{k,l \in \mathbb{Z}^2} c_{kl} e_k(x) e_l(y)$, where $e_k(x) = e^{2i\pi kx}$ and $\sum_{k,l} |c_{kl}|^2 < \infty$.

4.6.1 Poincaré, Birkhoff, and Ergodic Decomposition Theorems

The first important result from ergodic theory is the Poincaré recurrence theorem. Notice that there is no assumption here that $\mathbb{P}$ is ergodic.

Theorem 4.41 (Poincaré Recurrence Theorem) *Let $\mathbb{P}$ be a T-invariant probability measure. For every measurable set $A \subset X$,*

$$\mathbb{P}(A) = \mathbb{P}(\{x \in A : T^n(x) \in A \text{ for infinitely many } n\}).$$

Proof For $N \in \mathbb{N}$, let

$$B_N := \{x \in A : \{T^n(x) : n \geq N\} \subset X \setminus A\}.$$

Then $T^{-n}(B_1) \cap B_1 = \emptyset$ for all $n \geq 1$. Hence $T^{-n}(B_1) \cap T^{-m}(B_1) = \emptyset$ for all $m, n \in \mathbb{N}$ and $n \neq m$. Thus

$$1 \geq \sum_{n \in \mathbb{N}} \mathbb{P}(T^{-n}(B_1)) = \sum_{n \in \mathbb{N}} \mathbb{P}(B_1)$$

and $\mathbb{P}(B_1) = 0$. Replacing T with T^N proves that $\mathbb{P}(B_N) = 0$. $\qquad\qquad \square$

Let $\mathcal{I}$ denote the set of all invariant sets. Then $\mathcal{I}$ is a σ-field. The next result is the celebrated pointwise Birkhoff ergodic theorem. The proof given here follows [42] and goes back to Neveu.

Theorem 4.42 (Birkhoff Ergodic Theorem) *Let $\mathbb{P}$ be a T-invariant probability measure and let $f \in L^1(\mathbb{P})$. Then $\hat{f} := \mathbb{E}(f|\mathcal{I})$ is $(T, \mathbb{P})$-invariant and*

$$\lim_{n \to \infty} \frac{1}{n} \sum_{i=0}^{n-1} f \circ T^i = \hat{f}$$

$\mathbb{P}$-*almost surely. In particular, if $\mathbb{P}$ is T-ergodic, then*

$$\lim_{n \to \infty} \frac{1}{n} \sum_{i=0}^{n-1} f \circ T^i = \mathbb{E}(f)$$

$\mathbb{P}$-*almost surely.*

Proof For $f \in L^1(\mathbb{P})$, set $S_n(f)(x) := \sum_{i=0}^{n-1} f \circ T^i(x)$ and $\hat{f} := \mathbb{E}(f|\mathcal{I})$. We claim that

$$\hat{f} < 0, \quad \mathbb{P}\text{-almost surely} \implies \limsup_{n \to \infty} \frac{S_n(f)}{n} \leq 0, \quad \mathbb{P}\text{-almost surely}.$$

Let us first derive the theorem from the claim. For $\varepsilon > 0$, set $f_\varepsilon := f - \hat{f} - \varepsilon$. Then $\hat{f}_\varepsilon = -\varepsilon < 0$, and since $\hat{f}$ is $(T, \mathbb{P})$-invariant (the proof is easy and left to the reader),

$$\limsup_{n\to\infty} \frac{S_n(f)}{n} - \hat{f} - \varepsilon = \limsup_{n\to\infty} \frac{S_n(f_\varepsilon)}{n} \leq 0.$$

Thus, ε being arbitrary, $\limsup_{n\to\infty} \frac{S_n(f)}{n} \leq \hat{f}$. Similarly, $\liminf_{n\to\infty} \frac{S_n(f)}{n} \geq \hat{f}$.

We now move on to the proof of the claim. For $n \in \mathbb{N}^*$ and $x \in X$, let

$$F_n(x) := \max\{S_k(f)(x) : k = 1, \ldots, n\},$$

$F_\infty(x) := \lim_{n\to\infty} F_n(x) \in \mathbb{R} \cup \{\infty\}$, and $A := \{F_\infty = \infty\}$. Clearly

$$\limsup_{n\to\infty} \frac{S_n(f)}{n} \leq 0$$

on $X \setminus A$ and it suffices to prove that $\mathbb{P}(A) = 0$. Now observe that $F_{n+1} - F_n \circ T = f - \min(0, F_n \circ T)$. Consequently, $A \in \mathcal{I}$ and $(F_{n+1} - F_n \circ T)$ decreases to $f - \min(0, F_\infty \circ T)$. In particular, by monotone convergence, $\lim_{n\to\infty} \mathbb{E}((F_{n+1} - F_n \circ T)\mathbf{1}_A) = \mathbb{E}(f\mathbf{1}_A) = \mathbb{E}(\hat{f}\mathbf{1}_A)$. By T-invariance of $\mathbb{P}$, the left-hand side is nonnegative. Hence, if $\hat{f} < 0$, $\mathbb{P}$-almost surely, then necessarily $\mathbb{P}(A) = 0$. $\qquad\square$

The next theorem, known as the ergodic decomposition theorem, shows that every invariant measure on a Borel subset of a Polish space equipped with the Borel σ-field can be written as a "sum" of ergodic measures.

Theorem 4.43 (Ergodic Decomposition Theorem) *Let M be a Borel subset of a Polish space, with Borel σ-field $\mathcal{B}(M)$. Let $T : M \to M$ be a measurable transformation. Every T-invariant probability measure $\mathbb{P}$ can be decomposed as*

$$\mathbb{P}(\cdot) = \int_M P(x, \cdot)\, \mathbb{P}(dx),$$

where P is a Markov kernel on $(M, \mathcal{B}(M))$ such that $P(x, \cdot)$ is ergodic for $\mathbb{P}$-almost every x.

Before proving the ergodic decomposition theorem, we state without proof a lemma that can be deduced from Theorem 10.2.2 in [21] and the monotone class theorem in the appendix.

Lemma 4.44 *Let M be a Borel subset of a Polish space, with Borel σ-field $\mathcal{B}(M)$. Let $\mathbb{P}$ be a probability measure on $(M, \mathcal{B}(M))$, and let $\mathcal{A}$ be a sub-σ-field of $\mathcal{B}(M)$. Then there exists a Markov kernel P on $(M, \mathcal{B}(M))$ such that for every $f \in B(M)$, Pf is a representative of $\mathbb{E}(f|\mathcal{A})$, i.e., Pf is $\mathcal{A}$-measurable and $\mathbb{E}(\mathbf{1}_A Pf) = \mathbb{E}(\mathbf{1}_A f)$ for every $A \in \mathcal{A}$.*

Proof *(Theorem 4.43)* Recall that $\mathcal{I}$ denotes the σ-field of T-invariant sets in $\mathcal{B}(M)$. By Lemma 4.44, there is a Markov kernel P on $(M, \mathcal{B}(M))$ such that for every $f \in B(M)$, Pf is a representative of $\mathbb{E}(f|\mathcal{I})$. This yields

$$\mathbb{P}(A) = \mathbb{E}(\mathbb{E}(\mathbf{1}_A|\mathcal{I})) = \int_M P(x, A)\, \mathbb{P}(dx), \quad \forall A \in \mathcal{B}(M).$$

As a subset of a separable metric space, M is separable (see Exercise 4.45 (*ii*) below). Proposition 4.5 implies the existence of a countable family $\{f_n\}_{n \in \mathbb{N}} \subset C_b(M)$ such that for every $\mu, \nu \in \mathcal{P}(M)$, $\mu = \nu$ if and only if $\mu f_n = \nu f_n$ for all $n \in \mathbb{N}$. For every $n \in \mathbb{N}$, Pf_n is a representative of $\mathbb{E}(f_n|\mathcal{I})$, and $x \mapsto P(x, T^{-1}(\cdot))f_n = P(f_n \circ T)(x)$ is a representative of $\mathbb{E}(f_n \circ T|\mathcal{I})$. Since $\mathbb{P}$ is T-invariant, we have $\mathbb{E}(f_n|\mathcal{I}) = \mathbb{E}(f_n \circ T|\mathcal{I})$ for every $n \in \mathbb{N}$, hence $P(x, \cdot)$ is T-invariant for $\mathbb{P}$-almost every x.

To show that $P(x, \cdot)$ is ergodic for $\mathbb{P}$-almost every x, we follow the proof of Theorem 6.2 in [24]. Since M is a separable metric space, the σ-field $\mathcal{B}(M)$ is countably generated, i.e., there is a countable family of sets $\{A_n\}_{n \in \mathbb{N}}$ such that $\mathcal{B}(M) = \sigma(A_n : n \in \mathbb{N})$ (see Exercise 4.45 (*iii*)). As a result, $L^1(M, \mathcal{B}(M), \mathbb{P})$ is separable (see parts (*i*) and (*ii*) of Exercise 4.46 below). Since the set $\{\mathbf{1}_A : A \in \mathcal{I}\}$ is contained in $L^1(M, \mathcal{B}(M), \mathbb{P})$, it is also separable in the L^1-topology, so there is a countable family $\{A_n\}_{n \in \mathbb{N}} \subset \mathcal{I}$ such that for every $A \in \mathcal{I}$ and for every $\epsilon > 0$, there is $n \in \mathbb{N}$ with $\mathbb{P}(A \triangle A_n) < \epsilon$.

Let $\mathcal{I}_0 := \sigma(A_n : n \in \mathbb{N})$. By definition, $\mathcal{I}_0$ is a countably generated sub-σ-field of $\mathcal{I}$. Moreover, $\mathcal{I}_0$ and $\mathcal{I}$ are $\mathbb{P}$-equivalent, i.e., for every $A \in \mathcal{I}$ there is $B \in \mathcal{I}_0$ such that $\mathbb{P}(A \triangle B) = 0$ (see Exercise 4.46 (*iii*)). As $\mathcal{I}$ need not be countably generated (see Exercise 4.48 below), we will work with $\mathcal{I}_0$ in the remainder of the proof. Applying Lemma 4.44 to $\mathcal{I}_0$, we obtain a Markov kernel Q on $(M, \mathcal{B}(M))$ such that for every $f \in B(M)$, Qf is a representative of $\mathbb{E}(f|\mathcal{I}_0)$. Let $\{f_n\}_{n \in \mathbb{N}} \subset C_b(M)$ be as above. For $n \in \mathbb{N}$, consider the function $h_n := Qf_n$. Since $\mathcal{I}$ and $\mathcal{I}_0$ are $\mathbb{P}$-equivalent, $\mathbb{E}(f_n|\mathcal{I}) = \mathbb{E}(f_n|\mathcal{I}_0)$. As a result, there is $M^1 \in \mathcal{B}(M)$ such that $\mathbb{P}(M^1) = 1$ and for every $x \in M^1$, $P(x, \cdot)$ is T-invariant and

$$h_n(x) = P(x, \cdot)f_n, \quad \forall n \in \mathbb{N}.$$

Hence, $Q(x, \cdot) = P(x, \cdot)$ is T-invariant for every $x \in M^1$. By Birkhoff's ergodic theorem 4.42, there is $M^2 \subset M^1$ such that $\mathbb{P}(M^2) = 1$ and for every $x \in M^2$,

$$\lim_{N \to \infty} \frac{1}{N} \sum_{k=0}^{N-1} f_n(T^k(x)) = h_n(x), \quad \forall n \in \mathbb{N}.$$

And as both $Q(\cdot, A_n)$ and $\mathbf{1}_{A_n}$ are representatives of $\mathbb{E}(\mathbf{1}_{A_n}|\mathcal{I}_0)$, there is $M^3 \subset M^2$ such that $\mathbb{P}(M^3) = 1$ and $Q(x, A_n) = \mathbf{1}_{A_n}(x)$ for every $x \in M^3$ and $n \in \mathbb{N}$. Finally, as $Q(\cdot, M^3)$ is a representative of $\mathbb{E}(\mathbf{1}_{M^3}|\mathcal{I}_0)$ and as $\mathbb{P}(M^3) = 1$, there is $M^4 \subset M^3$ such that $\mathbb{P}(M^4) = 1$ and $Q(x, M^3) = 1$ for every $x \in M^4$.

Let us show that $Q(x, \cdot)$ is ergodic for every $x \in M^4$, which will complete the proof of the ergodic decomposition theorem. Fix $x \in M^4$ and $A \in \mathcal{I}$. In light of Proposition 4.36, it is enough to show that $Q(x, A) \in \{0, 1\}$. If $Q(x, A) = 0$, we are done. If $Q(x, A) > 0$, consider the probability measure

$$\nu(B) := \frac{Q(x, A \cap B)}{Q(x, A)}, \quad B \in \mathcal{B}(M).$$

Since $\nu(A) = 1$, it suffices to show that $Q(x, \cdot) = \nu$, which will follow from

$$h_n(x) = \nu f_n, \quad \forall n \in \mathbb{N}. \tag{4.13}$$

Set

$$[x] := \bigcap_{A \in \mathcal{I}_0 : x \in A} A.$$

By Exercise 4.47 (i) below, one has

$$[x] = \bigcap_{n : x \in A_n} A_n \cap \bigcap_{n : x \notin A_n} A_n^c \tag{4.14}$$

and $[x] \in \mathcal{I}_0$. Fix $n \in \mathbb{N}$. Since h_n is $\mathcal{I}_0$-measurable, it is constant on the set $[x]$ by Exercise 4.47 (ii). Therefore, we have for every $y \in [x] \cap M^3$

$$h_n(x) = h_n(y) = \lim_{N \to \infty} \sum_{k=0}^{N-1} f_n(T^k(y)).$$

Since $x \in M^3$, the representation of $[x]$ in (4.14) implies $Q(x, [x]) = 1$ and thus $Q(x, [x] \cap M^3) = 1$. Since $Q(x, \cdot)$ is T-invariant, another application of Birkhoff's ergodic theorem then yields that the constant $h_n(x)$ is a representative of $\mathsf{E}_{Q(x,\cdot)}(f_n|\mathcal{I})$, where $\mathsf{E}_{Q(x,\cdot)}$ denotes expectation with respect to $Q(x, \cdot)$. Consequently,

$$h_n(x)Q(x, A) = \mathsf{E}_{Q(x,\cdot)}(\mathbf{1}_A h_n(x)) = \mathsf{E}_{Q(x,\cdot)}(\mathbf{1}_A f_n) = \int_M \mathbf{1}_A(z) f_n(z) \, Q(x, dz).$$

Dividing both sides by $Q(x, A)$ gives (4.13). $\qquad\qquad\square$

Exercise 4.45 (Properties of Separable Metric Spaces) Let (M, d) be a separable metric space.

(i) Let $D \subset M$ be countable and dense. Show that $\{B(x, r) : x \in D, r \in \mathbb{Q}_+^*\}$ is a basis for the topology on M, where $B(x, r)$ stands for the open ball with center x and radius r.

(ii) Let A be any subset of M. Show that A with the metric induced from M is itself a separable metric space.
(iii) Show that the Borel σ-field $\mathcal{B}(M)$ is countably generated.

Exercise 4.46 For an arbitrary probability space $(\Omega, \mathcal{F}, \mathbb{P})$, prove the following statements:

(i) If $\mathcal{F}$ is countably generated, then $(\Omega, \mathcal{F}, \mathbb{P})$ is separable, i.e., there is a countable family $\mathcal{D} \subset \mathcal{F}$ such that for every $A \in \mathcal{F}$ and $\epsilon > 0$ there is $B \in \mathcal{D}$ with $\mathbb{P}(A \triangle B) < \epsilon$.
(ii) If $(\Omega, \mathcal{F}, \mathbb{P})$ is separable, then $L^1(\Omega, \mathcal{F}, \mathbb{P})$ is a separable metric space.
(iii) If $(\Omega, \mathcal{F}, \mathbb{P})$ is separable, then for every $A \in \mathcal{F}$ there is $B \in \sigma(\mathcal{D})$ such that $\mathbb{P}(A \triangle B) = 0$.

Exercise 4.47 Let $(\Omega, \mathcal{F})$ be a measurable space, let $\{A_n\}_{n \in \mathbb{N}} \subset \mathcal{F}$ be a countable family of sets, and let $\mathcal{A} := \sigma(A_n : n \in \mathbb{N})$. For $x \in \Omega$, set

$$[x]_\mathcal{A} := \bigcap_{A \in \mathcal{A} : x \in A} A.$$

(i) Show that for every $x \in \Omega$,

$$[x]_\mathcal{A} = \bigcap_{n : x \in A_n} A_n \cap \bigcap_{n : x \notin A_n} A_n^c,$$

and deduce that $[x]_\mathcal{A} \in \mathcal{A}$.
(ii) Let $f : \Omega \to \mathbb{R}$ be $\mathcal{A}$-measurable and let $x \in \Omega$. Show that f is constant on $[x]_\mathcal{A}$.

The next exercise shows that $\mathcal{I}$, the σ-field of T-invariant sets, need not be countably generated.

Exercise 4.48 Consider the irrational rotation T_α of Exercise 4.37 with α irrational. Let $\mathcal{I}$ be the σ-field of T_α-invariant sets. Use the formula from Exercise 4.47 (i) to show that $\mathcal{I}$ is not countably generated, even though $\mathcal{B}(S^1)$ is.

4.7 Application to Markov Chains

Consider now the canonical chain introduced in Sect. 1.3, Proposition 1.8. Let $\Theta : M^\mathbb{N} \to M^\mathbb{N}$ be the shift operator defined by $\Theta(\omega)_n := \omega_{n+1}$ and let $\mathbb{P}_\nu$ be the law of the canonical chain with initial distribution ν and kernel P. Recall that $\mathbb{P}_\nu$ is a probability measure over $M^\mathbb{N}$ characterized by (1.3).

Proposition 4.49

(i) $\mathbb{P}_v$ *is Θ-invariant if and only if $v \in \mathsf{Inv}(P)$.*

(ii) *Let $v \in \mathsf{Inv}(P)$ and let $h \in L^1(\mathbb{P}_v)$ be $(\Theta, \mathbb{P}_v)$-invariant. For $x \in M$ such that $h \in L^1(\mathbb{P}_x)$, let*

$$\overline{h}(x) := \mathbb{E}_x(h) = \int h \, d\mathbb{P}_x.$$

Then

(a) $h(\omega) = \overline{h}(\omega_0)$, $\mathbb{P}_v$*-almost surely;*

(b) $\overline{h}$ *is (P, v)-invariant.*

(iii) $\mathbb{P}_v$ *is Θ-ergodic if and only if v is P-ergodic.*

Proof

(i) This follows easily from the definitions.

(ii) Let $h \in L^1(\mathbb{P}_v)$ be $(\Theta, \mathbb{P}_v)$-invariant. For $n \in \mathbb{N}$, set $h_n := \mathbb{E}_v(h|\mathcal{F}_n)$. By Doob's martingale convergence theorem (Theorem A.7), h_n converges $\mathbb{P}_v$-almost surely, hence in probability, to h. In particular, for all $\varepsilon > 0$, $\lim_{n \to \infty} \mathbb{P}_v(|h_{n+1} - h_n| > \varepsilon) = 0$. By $(\Theta, \mathbb{P}_v)$-invariance of h and by the Markov property from Proposition 1.10,

$$h_n = \mathbb{E}_v(h \circ \Theta^n | \mathcal{F}_n) = \mathbb{E}_{\omega_n}(h) = \overline{h}(\omega_n).$$

Thus,

$$\mathbb{P}_v(|h_{n+1} - h_n| > \varepsilon) = \mathbb{P}_v(|\overline{h}(\omega_{n+1}) - \overline{h}(\omega_n)| > \varepsilon). \tag{4.15}$$

Since $v \in \mathsf{Inv}(P)$, (i) implies that $\mathbb{P}_v$ is Θ-invariant. The expression on the right-hand side of (4.15) thus equals $\mathbb{P}_v(|\overline{h}(\omega_1) - \overline{h}(\omega_0)| > \varepsilon)$, which proves that $h_n = h_0 = h$. Also, by the Markov property, $P\overline{h}(x) = \mathbb{E}_x(\mathbb{E}_{X_1}(h)) = \mathbb{E}_x(\mathbb{E}_x(h \circ \Theta | \mathcal{F}_1)) = \mathbb{E}_x(h \circ \Theta)$. And as h is $(\Theta, \mathbb{P}_v)$-invariant, we have for v-almost every $x \in M$ that $\mathbb{E}_x(h \circ \Theta) = \overline{h}(x)$.

(iii) Let v be P-ergodic. We will show that every $(\Theta, \mathbb{P}_v)$-invariant function $h \in L^1(\mathbb{P}_v)$ is $\mathbb{P}_v$-almost surely constant. In particular, every $(\Theta, \mathbb{P}_v)$-invariant set has $\mathbb{P}_v$-measure 0 or 1, so $\mathbb{P}_v$ is Θ-ergodic by Proposition 4.36. If $h \in L^1(\mathbb{P}_v)$ is $(\Theta, \mathbb{P}_v)$-invariant, then $\overline{h}$ is v-almost surely constant by (ii) and P-ergodicity of v. By (ii), this proves that h is $\mathbb{P}_v$-almost surely constant.

Conversely, assume that $\mathbb{P}_v$ is Θ-ergodic. Let A be a (P, v)-invariant set. Set $\tilde{A} := \{\omega \in M^{\mathbb{N}} : \omega_0 \in A\}$. Then $\mathbb{P}_v(\tilde{A} \cap \Theta^{-1}(\tilde{A})) = \int_A v(dx)P(x, A) = v(A) = \mathbb{P}_v(\tilde{A})$. This shows that $\tilde{A}$ is $(\Theta, \mathbb{P}_v)$-invariant. Hence $v(A) = \mathbb{P}_v(\tilde{A}) \in \{0, 1\}$. □

Theorem 4.50 *Let P be a Markov kernel, $\mu \in \mathsf{Inv}(P)$, and $h \in L^1(\mathbb{P}_\mu)$. Then there exist a set $N \in \mathcal{B}(M)$ and a function $\overline{h} \in L^1(\mu)$ such that $\mu(N) = 1$ and, for all $x \in N$,*

$$\lim_{n \to \infty} \frac{1}{n} \sum_{k=0}^{n-1} h \circ \Theta^k(\omega) = \overline{h}(x)$$

$\mathbb{P}_x$-almost surely. If μ is ergodic, then $\overline{h}(x) = \mathbb{E}_\mu(h)$.

Proof By Birkhoff's ergodic theorem, $\frac{1}{n} \sum_{k=0}^{n-1} h \circ \Theta^k(\omega)$ converges $\mathbb{P}_\mu$-almost surely to a $(\Theta, \mathbb{P}_\mu)$-invariant function $\hat{h} \in L^1(\mathbb{P}_\mu)$. According to Proposition 4.49 (ii), $\hat{h}(\omega) = \overline{h}(\omega_0)$, $\mathbb{P}_\mu$-almost surely, where $\overline{h}(\omega_0) := \mathbb{E}_{\omega_0}(h)$. To conclude the proof, we use the fact that $\mathbb{P}_\mu(\cdot) = \int_M \mathbb{P}_x(\cdot) \mu(dx)$. □

The next theorem is the ergocic decomposition theorem for a Markov kernel.

Theorem 4.51 *Let M be a Borel subset of a Polish space and let P be a Markov kernel on $(M, \mathcal{B}(M))$. Every P-invariant probability measure μ can be decomposed as*

$$\mu(\cdot) = \int_M Q(x, \cdot) \, \mu(dx), \tag{4.16}$$

where Q is a Markov kernel on $(M, \mathcal{B}(M))$ such that $Q(x, \cdot)$ is P-ergodic for μ-almost every x.

Proof Let $\mathcal{I}(P, \mu)$ be the collection of (P, μ)-invariant sets in $\mathcal{B}(M)$. In Exercise 4.52 below, you are asked to show that $\mathcal{I}(P, \mu)$ is a σ-field. By Lemma 4.44, there is a Markov kernel Q on $(M, \mathcal{B}(M))$ such that for every $f \in B(M)$, Qf is a representative of $\mathbb{E}_\mu(f | \mathcal{I}(P, \mu))$, where $\mathbb{E}_\mu$ denotes expectation with respect to μ. In complete analogy to the proof of Theorem 4.43, this yields the representation in (4.16).

It remains to show that $Q(x, \cdot)$ is P-ergodic for μ-almost every $x \in M$. Let $(\tilde{M}, d)$ be a Polish space such that M is a Borel subset of $\tilde{M}$. The space $\tilde{M}^{\mathbb{N}}$ equipped with the metric

$$e(\omega, \alpha) := \sum_{i \in \mathbb{N}} 2^{-i} \frac{d(\omega_i, \alpha_i)}{1 + d(\omega_i, \alpha_i)}$$

is Polish as well; the corresponding Borel σ-field equals the product σ-field $\mathcal{B}(\tilde{M})^{\otimes \mathbb{N}}$. Thus, $M^{\mathbb{N}}$ is a Borel subset of the Polish space $\tilde{M}^{\mathbb{N}}$. By Proposition 4.49 (i), the Markov measure $\mathbb{P}_\mu$ on $(M^{\mathbb{N}}, \mathcal{B}(M)^{\otimes \mathbb{N}})$ is Θ-invariant. Hence, by the ergodic decomposition theorem 4.43, there is a Markov kernel $\mathcal{P}$ on $(M^{\mathbb{N}}, \mathcal{B}(M)^{\otimes \mathbb{N}})$ such that

$$\mathbb{P}_\mu(\cdot) = \int_{M^{\mathbb{N}}} \mathcal{P}(\omega, \cdot)\, \mathbb{P}_\mu(d\omega),$$

and $\mathcal{P}(\omega, \cdot)$ is Θ-ergodic for $\mathbb{P}_\mu$-almost every $\omega \in M^{\mathbb{N}}$. Moreover, as seen in the proof of Theorem 4.43, $\mathcal{P}f$ is a representative of $\mathbb{E}_\mu(f|\mathcal{I})$ for every $f \in B(M^{\mathbb{N}})$, where $\mathcal{I}$ is the σ-field of Θ-invariant sets in $\mathcal{B}(M)^{\otimes \mathbb{N}}$. We will now relate the Markov kernels Q and $\mathcal{P}$ by showing that $\mathbb{P}_\mu$-almost surely,

$$\mathcal{P}(\omega, \cdot) = \mathbb{P}_{Q(\omega_0, \cdot)}(\cdot).$$

Let $\{F_n\}_{n \in \mathbb{N}} \subset C_b(M^{\mathbb{N}})$ such that for every $\mathsf{P}, \mathsf{Q} \in \mathcal{P}(M^{\mathbb{N}})$, $\mathsf{P} = \mathsf{Q}$ if and only if $\mathsf{P}F_n = \mathsf{Q}F_n$ for all $n \in \mathbb{N}$. In Exercise 1.9, we introduced the canonical projections $\pi_n : M^{\mathbb{N}} \to M^{n+1}$, $\omega \mapsto (\omega_i)_{i=0,\dots,n}$. We use π_0 to define the σ-field

$$\mathcal{J} := \{\pi_0^{-1}(A) : A \in \mathcal{I}(P, \mu)\}.$$

Claim: The σ-fields $\mathcal{J}$ and $\mathcal{I}$ are $\mathbb{P}_\mu$-equivalent.
Proof of the claim: Let $S \in \mathcal{I}$ and define $\varphi(x) := \mathbb{P}_x(S)$ and

$$A := \{x \in M : \varphi(x) = 1\}.$$

By Proposition 4.49 $(ii)(b)$, φ is (P, μ)-invariant. Since every $x \in A$ such that $\varphi(x) = P\varphi(x)$ satisfies $P(x, A) = 1$, it follows with Exercise 4.52 (i) that $A \in \mathcal{I}(P, \mu)$, and hence $\pi_0^{-1}(A) \in \mathcal{J}$. By Proposition 4.49 $(ii)(a)$,

$$\mathbf{1}_S(\omega) = \varphi(\omega_0)$$

for $\mathbb{P}_\mu$-almost every ω. In particular, $\varphi(\omega_0) \in \{0, 1\}$, $\mathbb{P}_\mu$-almost surely, so

$$\mathbf{1}_S = \mathbf{1}_{\pi_0^{-1}(A)}, \qquad \mathbb{P}_\mu\text{-a.s.}$$

Hence,

$$\mathbb{P}_\mu(S \triangle \pi_0^{-1}(A)) = 0.$$

Let us now fix a set $S \in \mathcal{J}$. Then there is $A \in \mathcal{I}(P, \mu)$ such that $S = \pi_0^{-1}(A)$. Set $\tilde{S} := A^{\mathbb{N}} = \bigcap_{n \in \mathbb{N}} \pi_n^{-1}(A^{n+1})$. A simple induction argument using $A \in \mathcal{I}(P, \mu)$ implies that $\mathbb{P}_\mu(\pi_n^{-1}(A^{n+1})) = \mu(A)$ for all $n \in \mathbb{N}$. Continuity of $\mathbb{P}_\mu$ from above yields

$$\mathbb{P}_\mu(\tilde{S}) = \mu(A) = \mathbb{P}_\mu(S).$$

Since $\tilde{S} \subset S$, it follows that $\mathbb{P}_\mu(S \triangle \tilde{S}) = 0$. And in the proof of Proposition 4.36, it was shown that for every $\tilde{S} \in \mathcal{I}(P, \mu)$ there is $\hat{S} \in \mathcal{I}$ such that $\mathbb{P}_\mu(\hat{S} \triangle \tilde{S}) = 0$. This completes the proof of the claim.

We now complete the proof of Theorem 4.51. Since $\mathcal{I}$ and $\mathcal{J}$ are $\mathbb{P}_\mu$-equivalent, we have for every $n \in \mathbb{N}$ that the representatives of $\mathbb{E}_\mu(F_n|\mathcal{I})$ and the representatives of $\mathbb{E}_\mu(F_n|\mathcal{J})$ are representatives of $\mathbb{E}_\mu(F_n|\sigma(\mathcal{I}, \mathcal{J}))$. The function $\mathcal{P}F_n$ is a representative of $\mathbb{E}_\mu(F_n|\mathcal{I})$ and thus also of $\mathbb{E}_\mu(F_n|\sigma(\mathcal{I}, \mathcal{J}))$. Let

$$\mathcal{F}_0 := \{\pi_0^{-1}(A) : A \in \mathcal{B}(M)\}.$$

For $n \in \mathbb{N}$, consider the functions

$$\overline{F}_n : M \to \mathbb{R}, \ x \mapsto \mathbb{E}_x(F_n)$$

and

$$G_n : M^{\mathbb{N}} \to \mathbb{R}, \ \omega \mapsto \overline{F}_n(\omega_0).$$

By the Markov property from Proposition 1.10, G_n is a representative of $\mathbb{E}_\mu(F_n|\mathcal{F}_0)$. As a result,

$$\mathbb{E}_\mu(F_n|\mathcal{J}) = \mathbb{E}_\mu(\mathbb{E}_\mu(F_n|\mathcal{F}_0)|\mathcal{J}) = \mathbb{E}_\mu(G_n|\mathcal{J}).$$

Next, observe that $\omega \mapsto Q\overline{F}_n(\omega_0)$ is a representative of $\mathbb{E}_\mu(G_n|\mathcal{J})$, and thus also of $\mathbb{E}_\mu(F_n|\mathcal{J})$ and $\mathbb{E}_\mu(F_n|\sigma(\mathcal{I}, \mathcal{J}))$. This shows that

$$1 = \mathbb{P}_\mu\left(\left\{\omega \in M^{\mathbb{N}} : \mathcal{P}F_n(\omega) = Q\overline{F}_n(\omega_0)\right\}\right)$$

$$= \mathbb{P}_\mu\left(\left\{\omega \in M^{\mathbb{N}} : \mathcal{P}(\omega, \cdot)F_n = \mathbb{P}_{Q(\omega_0, \cdot)}F_n\right\}\right),$$

and hence

$$\mathbb{P}_\mu\left(\left\{\omega \in M^{\mathbb{N}} : \mathcal{P}(\omega, \cdot) = \mathbb{P}_{Q(\omega_0, \cdot)}\right\}\right) = 1.$$

Let $S \in \mathcal{B}(M)^{\otimes \mathbb{N}}$ such that $\mathbb{P}_\mu(S) = 1$ and for every $\omega \in S$, $\mathcal{P}(\omega, \cdot) = \mathbb{P}_{Q(\omega_0, \cdot)}$ and $\mathcal{P}(\omega, \cdot)$ is Θ-ergodic. By Proposition 4.49 (iii), $Q(\omega_0, \cdot)$ is P-ergodic for every $\omega \in S$. Since $S \in \mathcal{B}(M)^{\otimes \mathbb{N}}$ and since π_0 is continuous, the set $\pi_0(S)$ is analytic (see Theorem 13.2.1 in [21]). Theorem 13.2.6 in [21] implies that there are $A, N \in \mathcal{B}(M)$ and $B \subset N$ such that $\mu(N) = 0$ and $\pi_0(S) = A \cup B$. It follows that

$$1 = \mathbb{P}_\mu(S) \leq \mathbb{P}_\mu(\pi_0^{-1}(A \cup N)) = \mu(A \cup N) \leq \mu(A) + \mu(N) = \mu(A),$$

which completes the proof. $\qquad\square$

Exercise 4.52 Let

$$\mathcal{I}(P, \mu) := \{A \in \mathcal{B}(M) : \mathbf{1}_A = P(\cdot, A) \ \mu\text{-a.s.}\}$$

be the collection of (P, μ)-invariant sets in $\mathcal{B}(M)$.

(i) Show that

$$\mathcal{I}(P, \mu) = \{A \in \mathcal{B}(M) : \mu(\{x \in A : P(x, A^c) > 0\}) = 0\}.$$

(ii) With the help of the representation in part (i), show that $\mathcal{I}(P, \mu)$ is a σ-field.

Exercise 4.53 (Skew Product Chains) Let M, N be two metric spaces and

$$T : M \times N \to N,$$

$$(x, y) \mapsto T_x(y)$$

a measurable map. Let (X_n) be an M-valued Markov chain defined on some filtered probability space $(\Omega, \mathcal{F}, \mathbb{F}, \mathsf{P})$ and let $Y_0 \in N$ be an $\mathcal{F}_0$-measurable random variable. Consider the stochastic process (Y_n) defined by

$$Y_{n+1} = T_{X_n}(Y_n).$$

(i) Show that (X_n, Y_n) is a Markov chain on $(\Omega, \mathcal{F}, \mathbb{F}, \mathsf{P})$.
(ii) Suppose $\mu \in \mathcal{P}(M)$ is an invariant probability measure for (X_n) and $\nu \in \mathcal{P}(M)$ is T_x-invariant for all $x \in M$. Show that $\mu \otimes \nu$ is invariant for (X_n, Y_n).
(iii) We suppose here that μ is the **unique** invariant probability measure of (X_n).

  (a) Give an example where ν is T_x-ergodic, but $\mu \otimes \nu$ is not.
  (b) (inspired by Lemma 2.1 in [29]) Suppose that $\mu \otimes \nu$ is ergodic for (X_n, Y_n) and that for all $x \in M$, T_x is 1-Lipschitz, i.e.,

$$d(T_x(y), T_x(z)) \le d(y, z)$$

  for all $x \in M$, $y, z \in N$. Show that for all $f \in L_b(M \times N)$, μ-almost all $x \in M$, and **all** $y \in \mathsf{supp}(\nu)$,

$$\mathbb{P}_{x,y}\left(\lim_{n \to \infty} \frac{1}{n} \sum_{k=1}^n f(X_k, Y_k) = (\mu \otimes \nu)(f) \right) = 1.$$

  Deduce that, if $\mathsf{supp}(\nu) = N$, then (X_n, Y_n) is uniquely ergodic.

(iv) Using (iii) show that the map defined in Exercise 4.40 is uniquely ergodic. Deduce that for all β irrational, the sequence $(n^2\beta)_{n\geq 1}$ is equidistributed on S^1. *Hint:* Choose $\beta = 2\alpha$. See [28], Corollaries 1.12 and 1.13.

Exercise 4.54 (Markov Rotations) With the notation of the preceding Exercise 4.53, we assume here that $M = \{1, \ldots, n\}$, $N = S^1$, (X_n) is a Markov chain on M whose transition probability matrix K is irreducible, and that for all $i \in M$, $T_i(y) = y + \alpha_i$ for some $\alpha_i \in S^1$.

A *circuit* for K is a sequence $(i_1, \ldots, i_d)$ of $d \geq 1$ distinct points such that $K(i_k, i_{k+1}) > 0$ for $k = 1, \ldots, d$ and $i_{d+1} = i_1$. The purpose of this exercise is to show that the chain (X_n, Y_n) is uniquely ergodic if and only if there exists a circuit $(i_1, \ldots, i_d)$ such that $\alpha_{i_1} + \ldots + \alpha_{i_d}$ is irrational.

(i) (preliminary) Let D be a diagonal matrix whose entries $\theta_1, \ldots, \theta_n$ are complex numbers having modulus 1. Consider the linear equation

$$Ku = Du \qquad (4.17)$$

with $u \in \mathbb{C}^n$. Assume that $u \in \mathbb{C}^n$ is a nonzero solution to (4.17). Show that:

(a) $|u_i| = |u_1|$ for $i = 1, \ldots, n$;
(b) $K_{ij} > 0 \Rightarrow u_j = \theta_i u_i$;
(c) For every circuit $(i_1, \ldots, i_d)$, $\theta_{i_1} \ldots \theta_{i_d} = 1$.

Prove that there exists a nonzero solution to (4.17) if and only if for every circuit $(i_1, \ldots, i_d)$, $\theta_{i_1} \ldots \theta_{i_d} = 1$.

(ii) Let μ be the unique invariant probability measure of (X_n) and $f = (f_1, \ldots, f_n) \in L^2(\mu \otimes \lambda)$. Set $f_j(x) = \sum_{k\in\mathbb{Z}} u_j(k)e^{2i\pi kx}$ with $\sum_{k\in\mathbb{Z}} |u_j(k)|^2 < \infty$. Show that $Pf = f$ if and only if $Ku(k) = D^k u(k)$ for all $k \in \mathbb{Z}$, where D is the diagonal matrix with entries $e^{2i\pi\alpha_1}, \ldots, e^{2i\pi\alpha_n}$ and $u(k) = (u_j(k))_{j=1,\ldots,n}$. Here P stands for the kernel of (X_n, Y_n).

(iii) Prove the desired result.

4.8 Continuous Time: Invariant Probabilities for Markov Processes

Let $\{P_t\}_{t\geq 0}$ be a Markov semigroup on M, as defined in Sect. 1.5. A probability measure $\mu \in \mathcal{P}(M)$ is called *invariant for* $\{P_t\}_{t\geq 0}$ if it is invariant for P_t, **for all** $t \geq 0$, i.e., $\mu P_t = \mu$, $\forall t \geq 0$. As shown by the following simple example, being invariant for some P_t is not sufficient to be invariant for $\{P_t\}_{t\geq 0}$.

Example 4.55 Consider the deterministic continuous-time rotation on $M = \mathbb{R}/\mathbb{Z}$, given by $X_t^x = (x + t) \mod 1$. The associated semigroup is given by $P_t f(x) =$

$f(X_t^x)$. Its unique invariant probability measure is the uniform measure on M. However, for all $k \in \mathbb{N}^*$ and $x \in M$, $\frac{1}{k}\sum_{i=0}^{k-1} \delta_{x+i/k}$ is invariant for $P_{1/k}$.

Nevertheless, existence of an invariant probability measure for some P_t always implies existence of some invariant probability measure for $\{P_t\}_{t\geq 0}$.

Proposition 4.56 *Suppose μ is an invariant probability measure for P_T for some $T > 0$. Then*

$$\mu_T := \frac{1}{T}\int_0^T \mu P_s(\cdot)\, ds$$

is invariant for $\{P_t\}_{t\geq 0}$.

Proof For all $r > 0$ and $f \in B(M)$,

$$\int_0^T \mu P_s P_r f\, ds = \int_0^T \mu P_{s+r} f\, ds = \int_r^{T+r} \mu P_s f\, ds = \int_0^T \mu P_s f\, ds,$$

where the last equality follows from the fact that, by P_T-invariance, the map $s \mapsto \mu P_s f$ is T-periodic. $\qquad\square$

We now introduce a Markov kernel whose invariant probability measures coincide with the invariant probability measures of $\{P_t\}$. This kernel is usually called the *1-resolvent* (or simply the *resolvent*) of $\{P_t\}_{t\geq 0}$. It is defined, for all $f \in B(M)$, as

$$Gf = \int_0^\infty e^{-t} P_t f\, dt. \tag{4.18}$$

Proposition 4.57 *A probability measure μ is invariant for G if and only if it is invariant for $\{P_t\}_{t\geq 0}$.*

Proof Suppose $\mu G = \mu$. Then, for all $f \in B(M)$ and $s \geq 0$,

$$\mu P_s f = \mu G P_s f = e^s \int_M \int_0^\infty e^{-(t+s)} P_{t+s} f(x)\, dt\, \mu(dx) = e^s \int_s^\infty e^{-r} \mu P_r f\, dr.$$

This shows, by a simple bootstrap argument, that $s \mapsto \mu P_s f$ is C^1 and that $\frac{d}{ds}\mu P_s f|_{s=0} = 0$. Thus

$$\frac{d}{dt}\mu P_t f = \frac{d}{ds}\mu P_{t+s} f|_{s=0} = \frac{d}{ds}\mu P_s(P_t f)|_{s=0} = 0.$$

This proves that $\mu P_s f = \mu f$. The converse statement is obvious. $\qquad\square$

One of the main interests of Proposition 4.57 is that it allows to extend easily certain notions introduced for discrete-time chains to continuous-time processes.

For instance, an invariant probability measure for $\{P_t\}_{t\geq 0}$ is *ergodic for* $\{P_t\}_{t\geq 0}$ if it is ergodic for the Markov kernel G, as defined in Sect. 4.4. With such a definition, the results of Sect. 4.4 as well as the ergodic decomposition theorem 4.51 apply. Another consequence is the continuous-time version of the ergodic theorem given below as Proposition 4.58. We first define the notion of a progressive process. A continuous-time process $(X_t)_{t\geq 0}$ defined on a filtered probability space $(\Omega, \mathcal{F}, \mathbb{F}, \mathsf{P})$ is called *progressively measurable (with respect to* $\mathbb{F}$), or simply *progressive*, if for all $t \geq 0$, the map $(s, \omega) \in [0, t] \times \Omega \mapsto X_s(\omega) \in M$ is measurable with respect to $\mathcal{B}([0, t]) \otimes \mathcal{F}_t$. A progressive process is obviously adapted. Conversely, an adapted process having right-continuous (or left-continuous) paths is progressive (see, e.g., [45] for a proof).

Proposition 4.58 *Suppose* $(X_t)_{t\geq 0}$ *is a progressive Markov process with semigroup* $\{P_t\}_{t\geq 0}$. *Let* $U_1, U_2, \dots$ *be a sequence of independent identically distributed random variables having an exponential distribution with parameter 1 and independent of* $(X_t)_{t\geq 0}$. *Set* $T_0 = 0$, $T_{n+1} = T_n + U_{n+1}$ *for* $n \geq 0$, *and* $Y_n = X_{T_n}$ *for* $n \geq 0$. *Then*

(i) *The process* (Y_n) *is a Markov chain with kernel* G;
(ii) *For all* $f \in B(M)$,

$$\lim_{t\to\infty} \frac{1}{t} \int_0^t f(X_s)\, ds - \frac{1}{[t]} \sum_{k=0}^{[t]-1} Gf(Y_k) = 0$$

almost surely, where $[t] := \max\{z \in \mathbb{Z} : z \leq t\}$;
(iii) *In particular, if* μ *is ergodic for* $\{P_t\}_{t\geq 0}$ *and* X_0 *is distributed according to* μ, *then*

$$\lim_{t\to\infty} \frac{1}{t} \int_0^t f(X_s)\, ds = \mu(f)$$

almost surely.

Proof

(i) Let $g, h_0, \dots, h_n \in B(M)$. Set $\Sigma_n = \{(t_1, \dots, t_n) \in \mathbb{R}_+^n : t_1 \leq t_2 \leq \dots \leq t_n\}$. By Fubini's theorem and the Markov property,

$$\mathsf{E}(g(Y_{n+1})h_0(Y_0)\dots h_n(Y_n))$$

$$= \int_{\Sigma_n} \left(\int_{\mathbb{R}_+} \mathsf{E}(g(X_{t_n+u})h_0(X_0)h_1(X_{t_1})\dots h_n(X_{t_n}))\, e^{-u}\, du \right) e^{-t_n}\, dt_1 \dots dt_n$$

$$= \int_{\Sigma_n} \left(\int_{\mathbb{R}_+} \mathsf{E}(P_u g(X_{t_n})h_0(X_0)h_1(X_{t_1})\dots h_n(X_{t_n}))e^{-u}\, du \right) e^{-t_n}\, dt_1 \dots dt_n$$

$$= \int_{\Sigma_n} \mathsf{E}(Gg(X_{t_n})h_0(X_0)h_1(X_{t_1})\ldots h_n(X_{t_n}))e^{-t_n}\, dt_1\ldots dt_n$$

$$= \mathsf{E}(Gg(Y_n)h_0(Y_0)\ldots h_n(Y_n)).$$

(ii) Fix $f \in B(M)$ and let $\mathbf{t} = (t_k)_{k\geq 1}$ be a deterministic increasing sequence of positive numbers such that $t_n \uparrow \infty$ and $\limsup_{n\to\infty} \frac{1}{n}\sum_{k=1}^n (t_{k+1}-t_k)^2 < \infty$. Let

$$M_n^{\mathbf{t}} = \int_0^{t_n} f(X_s)\, ds - \sum_{k=1}^n \int_0^{t_k-t_{k-1}} P_s f(X_{t_{k-1}})\, ds,$$

with the convention that $t_0 = 0$. Then the sequence $(M_n^{\mathbf{t}})_{n\geq 0}$ is a martingale with respect to $\{\mathcal{F}_{t_n}\}_{n\geq 0}$ such that $\langle M \rangle_{n+1} - \langle M \rangle_n \leq (t_{n+1}-t_n)^2 \|f\|_\infty^2$. Thus, by the strong law of large numbers for martingales (see Theorem A.8),

$$\lim_{n\to\infty} \frac{M_n^{\mathbf{t}}}{n} = 0$$

almost surely.

Let now $\Sigma_\infty = \{\mathbf{t} \in \mathbb{R}_+^{\mathbb{N}^*} : 0 \leq t_1 \leq t_2 \ldots\}$ be equipped with its Borel σ-field and let ν denote the law of $(T_n)_{n\geq 1}$. By what precedes, for ν-almost every $\mathbf{t} \in \Sigma_\infty$, one has $\lim_{n\to\infty} \frac{M_n^{\mathbf{t}}(\omega)}{n} = 0$ for P-almost every $\omega \in \Omega$. Thus, by Fubini's theorem, the convergence of $M_n^{\mathbf{t}}(\omega)/n$ to 0 holds for $\nu \otimes \mathsf{P}$-almost every $(\mathbf{t}, \omega) \in \Sigma_\infty \times \Omega$.

The sequence $(M_n')_{n\geq 0}$ defined as

$$M_n' = \sum_{k=1}^n \left(\int_0^{T_k-T_{k-1}} P_s f(Y_{k-1})\, ds - Gf(Y_{k-1}) \right)$$

is a martingale with respect to the filtration $\{\mathcal{G}_n\}_{n\geq 0}$, where $\mathcal{G}_n = \sigma((Y_k, T_k) : 0 \leq k \leq n)$. Hence, relying again on the strong law of large numbers for martingales, $\lim_{n\to\infty} M_n'/n = 0$ almost surely. Since $\lim_{n\to\infty} T_n/n = \mathsf{E}(T_1) = 1$ holds P-almost surely, the desired convergence follows.

(iii) If μ is ergodic for $\{P_t\}$ and $X_0 = Y_0$ has law μ, then

$$\lim_{n\to\infty} \frac{1}{n}\sum_{k=0}^{n-1} Gf(Y_k) = \mu Gf = \mu f$$

almost surely by application of Theorem 4.50.

$\square$

Notes

The proof of the ergodic decomposition theorem 4.51 for a Markov kernel is taken from unpublished lecture notes by Yuri Bakhtin [4].

Chapter 5
Irreducibility

This chapter discusses different versions of irreducibility. The first one, called
ξ-*irreducibility*, or simply *irreducibility*, is a purely measure-theoretic notion
which generalizes the definition given for countable state spaces. An important
characteristic of irreducible chains is that they have at most one invariant probability
measure. Another notion, this time topological, is that the chain is *indecomposable*,
in the sense that there exists at least one *accessible point*, i.e., a point whose
every neighborhood has a positive probability of being touched by the chain
regardless of the initial condition. Indecomposability is not a sufficient condition
to ensure unique ergodicity for a Feller chain but it is for a strong Feller chain.
Moreover, the accessible set provides valuable information about the support of
invariant probability measures. The final section of the chapter introduces a weaker
condition than the strong Feller condition due to Hairer and Mattingly, called
the *asymptotic strong Feller property* and studies the structure of the ergodic
measures for chains satisfying this condition. For an asymptotic strong Feller chain,
the ergodic measures have disjoint support. On a connected space, an invariant
probability measure with full support is necessarily unique.

5.1 Resolvent and ξ-Irreducibility

Given a (nonzero) Borel measure ξ on M, P is called ξ-*irreducible* if for every
Borel set $A \subset M$ and every $x \in M$

$$\xi(A) > 0 \Rightarrow \exists k \geq 0, \, P^k(x, A) > 0.$$

Equivalently,

$$\xi(A) > 0 \Rightarrow R_a(x, A) > 0,$$

© The Author(s), under exclusive license to Springer Nature Switzerland AG 2022   85
M. Benaïm, T. Hurth, *Markov Chains on Metric Spaces*, Universitext,
https://doi.org/10.1007/978-3-031-11822-7_5

where $R_a(.,.)$ is the *resolvent kernel* defined as

$$R_a(x, A) = (1 - a) \sum_{k \geq 0} a^k P^k(x, A)$$

for some $0 < a < 1$.

Remarks 5.1

(i) Let (X_n) be a Markov chain with kernel P and (Δ_n) a sequence of i.i.d. random variables independent from (X_n) having a geometric distribution with parameter a, i.e.,

$$P(\Delta_i = k) = a^k(1 - a), \ k \in \mathbb{N};$$

Then R_a is the kernel of the sampled chain $Y_n = X_{T_n}$ with

$$T_n = \sum_{i=1}^{n} \Delta_i;$$

(ii) P and R_a have the same invariant probability measures;
(iii) If P is ξ-irreducible, then **for all** $n \in \mathbb{N}, x \in M$, and $A \in \mathcal{B}(M)$ such that $\xi(A) > 0$ there exists $k \geq n$ such that $P^k(x, A) > 0$.

Exercise 5.2

(i) Check the assertions of the preceding remark.
(ii) Using the notation of Remark 5.1, show that for all $m \in \mathbb{N}^*$, T_m has a *negative bimomial* distribution with parameters (a, m), i.e.,

$$P(T_m = k) = \binom{k + m - 1}{m - 1} a^k(1 - a)^m$$

for all $k \in \mathbb{N}$. Let $Y_n^m = X_{T_{nm}}$. Show that $(Y_n^m)_n$ is a Markov chain with kernel R_a^m.

Example 5.3 (Doeblin Condition) Suppose that, for some nonzero measure ξ, $R_a(x, A) \geq \xi(A)$ for all $x \in M$ and $A \in \mathcal{B}(M)$. Then P is ξ-irreducible.

Example 5.4 (Countable Chains) If M is countable and P is irreducible in the usual sense (see Chap. 2), then it is ξ-irreducible for $\xi = \sum_x \delta_x$.

Theorem 5.5 *Suppose that P is ξ-irreducible. Then P admits at most one invariant probability measure.*

Proof The assumption implies that ξ is absolutely continuous with respect to every invariant probability measure, but since distinct ergodic measures are mutually singular (Proposition 4.29), there is at most one such probability measure. If M

is a Borel subset of a Polish space, the ergodic decomposition theorem 4.51 implies the result.

For a general M (which does not even have to be a metric space but just a measurable set), we cannot rely on ergodic decomposition but can proceed as follows. Let us first observe that any two invariant probability measures μ, ν are equivalent, i.e., their null sets coincide. Indeed, by Lemma 4.26, the singular part of ν with respect to μ is either 0 or a nonzero invariant measure. The latter case is impossible because ξ is absolutely continuous with respect to any invariant probability measure. Thus, $\nu = h\mu$ with $h \in L^1(\mu)$. As shown in the proof of Proposition 4.29 (ii), for all $a > 0$ the measure $\mu_a = (h \wedge a)\mu$ is also invariant. Thus $(a - h \wedge a)\mu$ is either 0 or invariant. In the first case, $\mu(\{h \geq a\}) = 1$. In the second case, $\mu(\{h \geq a\}) = 0$ because $(a - h \wedge a)\mu$ and μ, both being invariant, are equivalent. This proves that h is μ-almost surely constant. Thus $\mu = \nu$. $\qquad\square$

5.2 The Accessible Set

With the exception of a few particular cases (such as Examples 5.3 and 5.4) it is in general not an easy task to verify that a Markov chain is ξ-irreducible. A purely topological notion of irreducibility is defined below. Combined with the existence of certain points satisfying a local Doeblin condition (see Chap. 6), this will ensure ξ-irreducibility.

Recall that the (topological) *support* of a measure μ is the closed set $\mathsf{supp}(\mu)$ defined as the intersection of all closed sets $F \subset M$ such that $\mu(M \setminus F) = 0$. It enjoys the following properties:

(a) $\mu(M \setminus \mathsf{supp}(\mu)) = 0$;
(b) $x \in \mathsf{supp}(\mu)$ if and only if $\mu(O) > 0$ for every open set O containing x.

Exercise 5.6 Prove that assertions $(a), (b)$ above hold in any separable metric space. Use the fact that such a space has a countable basis of open sets (see Exercise 4.45 (i)).

We define the set of points that are *accessible* from $x \in M$ (for P) as

$$\Gamma_x = \mathsf{supp}(R_a(x, \cdot)).$$

Equivalently, y is accessible from x if for every neighborhood U of y there exists $k \geq 0$ such that $P^k(x, U) > 0$.

For $C \subset M$, we let $\Gamma_C = \cap_{x \in C} \Gamma_x$ denote the set of points that are accessible from C and $\Gamma := \Gamma_M$ the set of *accessible points*. Note that Γ_C is a closed (but possibly empty) set. We say that P is (topologically) *indecomposable* if $\Gamma \neq \emptyset$.

Remark 5.7 If P is ξ-irreducible, then it is indecomposable and

$$\mathsf{supp}(\xi) \subset \Gamma.$$

The converse implication is false in general (see Theorem 5.5 and Remark 5.10) but true for strong Feller chains (see Proposition 5.17).

Proposition 5.8 *Assume P is Feller and topologically indecomposable. Then*

(i) $P(x, \Gamma) = 1$ *for all $x \in \Gamma$;*
(ii) $\Gamma \subset \mathsf{supp}(\mu)$ *for all $\mu \in \mathsf{Inv}(P)$;*
(iii) *If Γ has nonempty interior, $\mathsf{supp}(\mu) = \Gamma$ for all $\mu \in \mathsf{Inv}(P)$;*
(iv) *If Γ is compact, there exists $\mu \in \mathsf{Inv}(P)$ such that $\mathsf{supp}(\mu) = \Gamma$;*
(v) *If Γ is compact and $g : \Gamma \to \mathbb{R}$ is a continuous and harmonic function on Γ (i.e., $Pg(x) = g(x)$ for all $x \in \Gamma$), then g is constant.*

Proof

(i) Let $x \in \Gamma$. It is enough to prove that $\mathsf{supp}(P(x, .)) \subset \Gamma$. Let $x^* \in \mathsf{supp}(P(x, .))$ and O an open set containing x^*. Then $\delta := P(x, O) > 0$. By Feller continuity and the Portmanteau theorem 4.1, $V := \{y \in M : P(y, O) > \delta/2\}$ is an open set containing x. Let $z \in M$ and $k \in \mathbb{N}$ be such that $P^k(z, V) > 0$ (recall that $x \in \Gamma$). Then

$$P^{k+1}(z, O) \geq \int_V P^k(z, dy)P(y, O) \geq \frac{\delta}{2}P^k(z, V) > 0.$$

This proves that $x^* \in \Gamma$.

(ii) Let $x \in \Gamma, U$ a neighborhood of x, and μ an invariant probability measure. Then $\mu(U) = \int \mu(dy)R(y, U) > 0$.

(iii) By invariance, $\mu(\Gamma) = \int_\Gamma \mu(dx)R(x, \Gamma) + \int_{\Gamma^c} \mu(dx)R(x, \Gamma)$, and since, by (i), $R(x, \Gamma) = 1$ for all $x \in \Gamma$, it follows that $\int_{\Gamma^c} \mu(dx)R(x, \Gamma) = 0$. If furthermore Γ has nonempty interior, then $R(x, \Gamma) > 0$ for all x, so that $\mu(\Gamma^c) = 0$. This proves that $\mathsf{supp}(\mu) \subset \Gamma$.

(iv) By (i), Feller continuity, and Theorem 4.20, there exists an invariant probability measure μ with $\mu(\Gamma) = 1$; hence the result.

(v) By (i) we can assume without loss of generality that $\Gamma = M$. By compactness, accessibility, and Feller continuity, for every open set $O \subset M$ there exists a finite cover of M by open sets $U_1, \ldots, U_k$, integers $n_1, \ldots, n_k$, and $\delta > 0$ such that $P^{n_i}(x, O) \geq \delta$ for all $x \in U_i, 1 \leq i \leq k$. Thus $\mathbb{P}_x(\tau_O > n) \leq (1 - \delta)$ for $n = \max(n_1, \ldots, n_k)$, hence $\mathbb{P}_x(\tau_O > kn) \leq (1 - \delta)^k$ by the Markov property. Thus $\mathbb{P}_x(\tau_O < \infty) = 1$. The assumption that g is harmonic makes $(g(X_n))$ a bounded martingale. It then converges $\mathbb{P}_x$-almost surely. If g is nonconstant, there exist $a < b$ such that $\{g < a\}$ and $\{g > b\}$ are nonempty open sets, and, by what precedes, (X_n) visits infinitely often these sets $\mathbb{P}_x$-almost surely, a contradiction.

□

Remark 5.9 The inclusion $\Gamma \subset \mathsf{supp}(\mu)$ does not require Feller continuity.

Remark 5.10 The inclusion $\Gamma \subset \mathsf{supp}(\mu)$ may be strict when Γ has empty interior as shown by the following exercise. Other examples where the inclusion $\Gamma \subset \mathsf{supp}(\mu)$ is strict can be found in [8] and [9].

Exercise 5.11 Let $F : \{0, 1\} \times [0, 1] \to [0, 1]$ be the map defined by

$$F(0, x) = ax, \ F(1, x) = bx(1 - x),$$

where $0 < a < 1$ and $1 < b < 4$. Let (X_n) be the Markov chain on $[0, 1]$ defined by $X_{n+1} = F(\theta_{n+1}, X_n)$, $X_0 = x > 0$, where (θ_n) is an i.i.d. Bernoulli sequence with distribution $(1 - p)\delta_0 + p\delta_1$ for some $0 < p < 1$. Show that $\Gamma = \{0\}$ and that when $(1 - p) \log a + p \log b > 0$, there exists an invariant probability measure μ such that $\mu(\{0\}) = 0$, hence $\mathsf{supp}(\mu) \not\subset \Gamma$.

In case P is uniquely ergodic on a compact set, it is topologically indecomposable.

Proposition 5.12 *Suppose M is compact, P is Feller and uniquely ergodic with* $\mathsf{Inv}(P) = \{\mu\}$. *Then P is indecomposable and* $\Gamma = \mathsf{supp}(\mu)$.

Proof By Proposition 5.8 it suffices to prove that Γ is nonempty. By Theorem 4.20, $\frac{1}{n} \sum_{k=1}^{n} P^k(x, \cdot) \Rightarrow \mu$ for all $x \in M$. Hence, for any open set O such that $\mu(O) > 0$, $\liminf_{n \to \infty} \frac{1}{n} \sum_{k=1}^{n} P^k(x, O) > 0$. Thus $R(x, O) > 0$. $\qquad\square$

A partial converse to Proposition 5.12 is the following result. Recall that $L_b(M)$ is the set of real-valued bounded Lipschitz functions on M.

Proposition 5.13 *Assume that M is compact, P is Feller, Γ has nonempty interior, and for all $f \in L_b(M)$ the sequence $(P^n f)_{n \geq 1}$ is equicontinuous. Then P is uniquely ergodic.*

Proof By equicontinuity of $(P^n f)_{n \geq 1}$, the sequence $(\overline{f}_n)_{n \geq 1}$ defined by

$$\overline{f}_n = \frac{\sum_{k=1}^{n} P^k f}{n}$$

is also equicontinuous, hence relatively compact in $C_b(M)$ by the Arzelà-Ascoli theorem. Let g be a limit point of $(\overline{f}_n)_{n \geq 1}$. Then g is continuous and $Pg = g$. By Proposition 5.8 (v), $g|_{\Gamma}$ is a constant C_f. Let now μ and ν be two invariant probability measures. Then $\mu Pf = \mu f$ implies that $\mu(\overline{f}_n) = \mu(f)$. Therefore $\mu(f) = \mu(g) = \mu(g|_{\Gamma}) = C_f$. Similarly $\nu(f) = C_f$. This proves that $\mu = \nu$. $\qquad\square$

Exercise 5.14 Deduce from Proposition 5.13 that the irrational rotation T_α (see Exercise 4.37) is uniquely ergodic.

Exercise 5.15 Let M be a compact space. Using the notation of Chap. 3 and Sect. 4.5.1, consider the Markov chain on M recursively defined by

$$X_{n+1} = F_{\theta_{n+1}}(X_n).$$

Assume that Θ is a metric space, $(\theta, x) \mapsto F_\theta(x)$ is continuous, and for each $\theta \in \Theta$, F_θ is Lipschitz with Lipschitz constant l_θ. Assume furthermore that

(i) $\int l_\theta\, m(d\theta) \leq 1$ (compare with the condition of Theorem 4.31);
(ii) For every $x \in M$ and every open set $O \subset M$, there exists a sequence $\theta_1, \ldots, \theta_n$ with $\theta_i \in \mathsf{supp}(m)$, $1 \leq i \leq n$, such that $f_{\theta_n} \circ \ldots f_{\theta_1}(x) \in O$.

Show that (X_n) is uniquely ergodic.

Remark 5.16 It is important to emphasize here that the condition that Γ has nonempty interior is not sufficient to ensure uniqueness of the invariant probability measure. For instance, Furstenberg, in a remarkable work [29] (see also [48]), has shown that for a convenient choice of $\alpha \in \mathbb{R}\backslash\mathbb{Q}$ and β a smooth map on $S^1 := \mathbb{R}/\mathbb{Z}$, the diffeomorphism

$$T : S^1 \times S^1 \to S^1 \times S^1,$$

$$(x, y) \mapsto (x + \alpha, y + \beta(x))$$

is minimal (i.e., all the orbits are dense) but not uniquely ergodic.

Another example is given by the Ising model on $\mathbb{Z}^2$. This is a Feller Markov process on the compact set $M = \{-1, 1\}^{\mathbb{Z}^2}$ for which all points are accessible (i.e., $\Gamma = M$) and which admits (at low temperature) several invariant probability measures. See Example 2.3 in [33] for a discussion and further references.

Recall that a function $f : M \to \mathbb{R}$ is *lower semicontinuous* (respectively, *upper semicontinuous*) at a point $x_0 \in M$ if

$$f(x_0) \leq \liminf_{x \to x_0} f(x), \quad \text{resp. } f(x_0) \geq \limsup_{x \to x_0} f(x).$$

Clearly, f is continuous at a point $x_0 \in X$ if and only if f is both upper and lower semicontinuous at x_0.

Proposition 5.17 *Suppose that P is topologically indecomposable and that for some $x^* \in \Gamma$ and all $A \in \mathcal{B}(M)$, $x \mapsto P(x, A)$ is lower semicontinuous at x^*. Then P is ξ-irreducible for $\xi = P(x^*, .)$. In particular P admits at most one invariant probability measure.*

Proof Let A be such that $P(x^*, A) > 0$. Then for all $x \in M$ there exist a neighborhood O of x^* and $n \geq 0$ such that $P^n(x, O) > 0$ and $P(y, A) > 0$ for all $y \in O$ (by lower semicontinuity of $x \mapsto P(x, A)$ at x^*). Thus $P^{n+1}\mathbf{1}_A(x) \geq \int_O P^n(x, dy)P(y, A) > 0$. $\square$

Note that the assumption that $x \mapsto P(x, A)$ is lower semicontinuous at x^* is automatically satisfied if P is strong Feller. Hence Proposition 5.17 gives a practical tool to ensure that a strong Feller chain is uniquely ergodic. Another result about strong Feller chains is the following.

Proposition 5.18 *Suppose that P is strong Feller. Then*

(**i**) *Two distinct ergodic measures have disjoint support;*
(**ii**) *The support of an invariant non-ergodic probability measure is disconnected;*
(**iii**) *If M is connected and P has an invariant probability measure having full support, then P is uniquely ergodic.*

Proof

(*i*) Let μ, ν be two distinct ergodic measures. By Proposition 4.29 they are mutually singular. Hence there exists a Borel set $A \subset M$ such that $\mu(A) = 1$ and $\nu(A) = 0$. The set $\{x \in M : P(x, A) = 1\}$ is closed (strong Feller property) and has μ-measure 1 because $1 = \mu(A) = \int \mu(dx) P(x, A)$. Thus $\mathrm{supp}(\mu) \subset \{x \in M : P(x, A) = 1\}$. Similarly $\mathrm{supp}(\nu) \subset \{x \in M : P(x, M \setminus A) = 1\}$.

(*ii*) Let μ be invariant and let A be such that $P\mathbf{1}_A = \mathbf{1}_A$, μ-almost surely, and $0 < \mu(A) < 1$. Set $f = P\mathbf{1}_A$. Then $f(x) \in \{0, 1\}$ for μ-almost every x and, by the strong Feller property, f is continuous. Thus f restricted to $\mathrm{supp}(\mu)$ takes values in $\{0, 1\}$. If now $\mathrm{supp}(\mu)$ is connected, then f restricted to $\mathrm{supp}(\mu)$ is constant and $\mu(A) \in \{0, 1\}$. (*iii*) follows from (*ii*).

$\square$

5.2.1 Continuous Time: Accessibility

For a continuous-time semigroup $\{P_t\}_{t \geq 0}$ one defines, by analogy with the discrete-time setting, the set of points that are accessible from $x \in M$ (for $\{P_t\}_{t \geq 0}$) as

$$\Gamma_x = \mathrm{supp}(G(x, \cdot)),$$

where G is the 1-resolvent (see Eq. (4.18)).

Proposition 5.19 *Suppose $\{P_t\}_{t \geq 0}$ is weakly Feller and let $x, y \in M$. Then the following assertions are equivalent:*

(**i**) *The point y is accessible from x for $\{P_t\}_{t \geq 0}$;*
(**ii**) *The point y is accessible from x for G;*
(**iii**) *For every neighborhood U of y there exists $t \geq 0$ such that $P_t(x, U) > 0$.*

Proof Clearly $(i) \Rightarrow (ii)$ and $(ii) \Rightarrow (iii)$ because

$$G^k(f) = \int_0^\infty \gamma_k(t) P_t f \, dt,$$

where $\gamma_k(t) = e^{-t} t^{k-1}/k!$. To prove that $(iii) \Rightarrow (i)$ suppose that $P_t(x, U) > 0$ for some $t \geq 0$. By the weak Feller property, $P_{s+t}(x, \cdot) \Rightarrow P_t(x, \cdot)$ as $s \downarrow 0$. Thus, by the Portmanteau theorem 4.1, $\liminf_{s \downarrow 0} P_{t+s}(x, U) > 0$. Hence $G(x, U) > 0$. $\square$

5.3 The Asymptotic Strong Feller Property

The asymptotic strong Feller property was introduced in [34] by Hairer and Mattingly to prove uniqueness for the invariant probability measure of the Navier–Stokes equation on the two-dimensional torus, subject to degenerate stochastic forcing. Before we define this property, we introduce some notation.

Let (M, d^*) be a separable metric space, with $\mathcal{P}(M)$ the space of probability measures on $(M, \mathcal{B}(M))$. One important idea in this section is to consider a whole family of metrics on M, but throughout, d^* will be the metric that gives rise to the topology on M, and in particular induces the σ-field $\mathcal{B}(M)$.

For any bounded metric d on M, we let $\mathrm{Lip}_1(d)$ denote the set of $\mathcal{B}(M)$-measurable functions $\phi : M \to \mathbb{R}$ such that

$$|\phi(x) - \phi(y)| \leq d(x, y), \quad \forall x, y \in M.$$

Notice that $\mathrm{Lip}_1(d)$ contains all constant functions. If the metric d is continuous with respect to the topology induced by d^* and if $\mathcal{B}_d(M)$ denotes the Borel σ-field with respect to d, then $\mathrm{Lip}_1(d)$ is equal to the set of $\mathcal{B}_d(M)$-measurable functions $\phi : M \to \mathbb{R}$ such that $|\phi(x) - \phi(y)| \leq d(x, y)$ for all $x, y \in M$. For $\mu, \nu \in \mathcal{P}(M)$, we define

$$\|\mu - \nu\|_d := \sup_{\phi \in \mathrm{Lip}_1(d)} (\mu\phi - \nu\phi).$$

Boundedness of d guarantees that every function in $\mathrm{Lip}_1(d)$ is bounded and thus integrable with respect to any Borel probability measure on M.

Exercise 5.20 Let d^* be bounded. Show that $(\mu, \nu) \mapsto \|\mu - \nu\|_{d^*}$ defines a bounded metric on $\mathcal{P}(M)$.

Remark 5.21 If $\delta(x, y) := \mathbf{1}_{x \neq y}$ is the discrete metric, then

$$\|\mu - \nu\|_\delta = \tfrac{1}{2}|\mu - \nu| := \tfrac{1}{2} \sup\{|\mu f - \nu f| : f \in B(M), \|f\|_\infty \leq 1\},$$

where $|\mu - \nu|$ is the so-called total variation distance between μ and ν. The latter will play a key role in Chap. 8.

We call a metric d on M *continuous* if it is continuous as a function from $M \times M$ to $[0, \infty)$, where $M \times M$ has the topology induced by the product metric $(d^* \star d^*)((x, y), (x', y')) := d^*(x, x') + d^*(y, y')$. Notice in particular that d^* itself is continuous. A sequence of metrics $(d_n)_{n \geq 1}$ on M is called *nondecreasing* if for every $n \in \mathbb{N}^*$,

$$d_{n+1}(x, y) \geq d_n(x, y), \quad \forall x, y \in M.$$

Recall that $\delta(x, y) := \mathbf{1}_{x \neq y}$ and that δ_x is the Dirac measure that assigns mass 1 to $\{x\}$.

Definition 5.22 (Hairer, Mattingly) We say that a Markov kernel P on M is asymptotically strong Feller at $x \in M$ if there exist a nondecreasing sequence $(n_k)_{k \geq 1}$ of positive integers and a nondecreasing sequence $(d_k)_{k \geq 1}$ of continuous metrics on M such that

$$\lim_{k \to \infty} d_k(y, z) = \delta(y, z), \quad \forall y, z \in M,$$

and

$$\inf \left\{ \limsup_{k \to \infty} \sup_{y \in U} \| \delta_x P^{n_k} - \delta_y P^{n_k} \|_{d_k} : U \text{ open}, x \in U \right\} = 0.$$

We call P asymptotically strong Feller if it is asymptotically strong Feller at every $x \in M$.

Since $(d_k)_{k \geq 1}$ is nondecreasing and converges to a bounded metric, each metric d_k is, of course, bounded.

5.3.1 Strong Feller Implies Asymptotic Strong Feller

In this subsection, we show that every strong Feller Markov kernel also has the asymptotic strong Feller property. The proof of this statement makes use of the ultra Feller property, which we now define. A Markov kernel P on M is called *ultra Feller* if the mapping $x \mapsto \delta_x P$ is continuous with respect to the total variation distance (see Remark 5.21). In particular, every ultra Feller Markov kernel is strong Feller. The following statement corresponds to Theorem 1.6.6 in [32]. It is due to Dellacherie and Meyer, see [18].

Proposition 5.23 Let P and Q be strong Feller Markov kernels on M. Then the Markov kernel PQ is ultra Feller.

The proof of Proposition 5.23 we present here is taken from [32]. It is an adaptation of an argument due to Seidler. We begin by stating two lemmas.

Lemma 5.24 *Let P be a strong Feller Markov kernel on M. Then there exists $\pi \in \mathcal{P}(M)$ such that $P(x, \cdot) \ll \pi$ for every $x \in M$.*

Proof Since M is separable, there is a dense sequence $(x_n)_{n \geq 1}$ of elements of M. We define the probability measure

$$\pi(A) := \sum_{n=1}^{\infty} 2^{-n} P(x_n, A), \quad A \in \mathcal{B}(M).$$

To obtain a contradiction, assume there is $x \in M$ such that $P(x, \cdot)$ is not absolutely continuous with respect to π. Then there is $A \in \mathcal{B}(M)$ such that $\pi(A) = 0$ and $P(x, A) > 0$. Let $f := 1_A \in \mathcal{B}(M)$. Since P is strong Feller, Pf is continuous. We have $Pf(x) = P(x, A) > 0$. Since $\pi(A) = 0$, we have $0 = P(x_n, A) = Pf(x_n)$ for every $n \in \mathbb{N}^*$. But then continuity of Pf and the fact that (x_n) is dense in M imply that $Pf \equiv 0$, a contradiction. $\square$

The following real-analysis lemma corresponds to Corollary 1.6.3 in [32]. Recall from the proof of Lemma 4.44 in Sect. 4.6 that a σ-field $\mathcal{F}$ is called *countably generated* if there exists a countable family of sets $\{A_n\}_{n \in \mathbb{N}}$ such that $\mathcal{F} = \sigma(A_n : n \in \mathbb{N})$.

Lemma 5.25 *Let $(\Omega, \mathcal{F}, \pi)$ be a measure space such that $\mathcal{F}$ is countably generated. Let (ϕ_n) be a bounded sequence in $L^\infty(\Omega, \mathcal{F}, \pi)$. Then there exist a subsequence $(\phi_{n_k})_{k \geq 1}$ and $\phi \in L^\infty(\Omega, \mathcal{F}, \pi)$ such that*

$$\lim_{k \to \infty} \int_\Omega \phi_{n_k}(x) f(x) \, \pi(dx) = \int_\Omega \phi(x) f(x) \, \pi(dx), \quad \forall f \in L^1(\Omega, \mathcal{F}, \pi).$$

Proof The space $L^\infty(\Omega, \mathcal{F}, \pi)$ being the dual of $L^1(\Omega, \mathcal{F}, \pi)$, its unit ball is compact for the weak* topology by the Banach–Alaoglu theorem. Furthermore, the assumption that $\mathcal{F}$ is countably generated makes $L^1(\Omega, \mathcal{F}, \pi)$ separable (see Exercise 4.46). Thus, the unit ball of $L^1(\Omega, \mathcal{F}, \pi)$ is sequentially compact for the weak* topology. This proves the result. $\square$

We proceed to the proof of Proposition 5.23.

Proof *(Proposition 5.23)* Since Q is strong Feller, Lemma 5.24 yields existence of a probability measure π on $(M, \mathcal{B}(M))$ such that $Q(x, \cdot) \ll \pi$ for every $x \in M$. To obtain a contradiction, suppose that the kernel PQ is not ultra Feller. Then there are $x \in M$ and $\varepsilon > 0$ such that for every open neighborhood U of x,

$$\sup_{y \in U} \|\delta_x PQ - \delta_y PQ\|_\delta > \varepsilon.$$

For $r > 0$ and $y \in M$, let $B_r(y) := \{z \in M : d^*(y, z) < r\}$ be the open d^*-ball of radius r centered at y. Then for every $n \in \mathbb{N}^*$ there is $y_n \in B_{1/n}(x)$ such that

$$\|\delta_x PQ - \delta_{y_n} PQ\|_\delta > \varepsilon.$$

According to Remark 5.21,

$$\sup_{\phi \in B(M):\|\phi\|_\infty \leq 1} (PQ\phi(x) - PQ\phi(y_n)) > 2\varepsilon, \quad \forall n \in \mathbb{N}^*,$$

where the expression on the left-hand side denotes the total variation distance between $\delta_x PQ$ and $\delta_{y_n} PQ$. As a result, there is a sequence $(\phi_n)_{n \geq 1}$ in $B(M)$ such that $\|\phi_n\|_\infty \leq 1$ and

$$PQ\phi_n(x) - PQ\phi_n(y_n) > 2\varepsilon, \quad \forall n \in \mathbb{N}^*. \tag{5.1}$$

Since M is a separable metric space, Exercise 4.45 (ii) implies that the σ-field $\mathcal{B}(M)$ is countably generated. And since (ϕ_n) is a bounded sequence in $L^\infty(M, \mathcal{B}(M), \pi)$, Lemma 5.25 implies that there exist a subsequence $(\phi_{n_k})_{k \geq 1}$ and a function $\phi \in L^\infty(M, \mathcal{B}(M), \pi)$ such that

$$\lim_{k \to \infty} \int_M \phi_{n_k}(x) f(x)\, \pi(dx) = \int_M \phi(x) f(x)\, \pi(dx), \quad \forall f \in L^1(M, \mathcal{B}(M), \pi).$$

Since $Q(x, \cdot) \ll \pi$ for every $x \in M$, we have that for every $x \in M$ there is $h_x \in L^1(M, \mathcal{B}(M), \pi)$ with $Q(x, dy) = h_x(y)\, \pi(dy)$. Then, for every $x \in M$,

$$\lim_{k \to \infty} Q\phi_{n_k}(x) = Q\phi(x).$$

To keep notation short, set $\psi_k := Q\phi_{n_k}$ for every $k \in \mathbb{N}^*$, and set $\psi := Q\phi$. We also introduce the functions $(\rho_j)_{j \geq 1}$ defined by

$$\rho_j(x) := \sup_{k \geq j} |\psi_k(x) - \psi(x)|, \quad x \in M,$$

and note that $\lim_{j \to \infty} \rho_j(x) = 0$ for every $x \in M$. For every $k \geq 1$,

$$\|\psi_k\|_\infty \leq \|\phi_{n_k}\|_\infty \leq 1 \quad \text{and} \quad \|\rho_k\|_\infty \leq \|\psi\|_\infty + \sup_{l \geq 1} \|\psi_l\|_\infty \leq \|\phi\|_\infty + 1,$$

so bounded convergence implies that

$$\lim_{k \to \infty} P\psi_k(x) = P\psi(x) \tag{5.2}$$

and

$$\lim_{j\to\infty} P\rho_j(x) = 0$$

for every $x \in M$. For every $m \in \mathbb{N}^*$,

$$\limsup_{j\to\infty} P\rho_j(y_{n_j}) \leq \limsup_{j\to\infty} P\rho_m(y_{n_j}) = P\rho_m(x)$$

because (ρ_j) is a nonincreasing sequence of nonnegative functions in $B(M)$, $\lim_{j\to\infty} y_{n_j} = x$, and P is strong Feller. Since the estimate above holds for every $m \in \mathbb{N}^*$ and since $\lim_{m\to\infty} P\rho_m(x) = 0$, it follows that

$$\lim_{j\to\infty} P\rho_j(y_{n_j}) = 0. \qquad (5.3)$$

Consequently,

$$\limsup_{k\to\infty} \left(PQ\phi_{n_k}(x) - PQ\phi_{n_k}(y_{n_k}) \right)$$

$$\leq \limsup_{k\to\infty} |P\psi_k(x) - P\psi(x)|$$

$$+ \limsup_{k\to\infty} |P\psi(x) - P\psi(y_{n_k})| + \limsup_{k\to\infty} |P\psi(y_{n_k}) - P\psi_k(y_{n_k})|$$

$$\leq \limsup_{k\to\infty} P\rho_k(y_{n_k}) = 0,$$

where we used (5.2), the assumption that P is strong Feller, and (5.3). This contradicts (5.1).   □

We are now ready to state and prove the main result of this subsection.

Proposition 5.26 *Let P be a Markov kernel on a separable metric space (M, d^*). If P is strong Feller, then it is also asymptotically strong Feller.*

Proof Consider the sequence of continuous metrics

$$d_k(x, y) := 1 \wedge (kd^*(x, y)), \quad k \in \mathbb{N}^*,$$

where $a \wedge b$ denotes the minimum of a and b. The sequence is clearly nondecreasing, and

$$\lim_{k\to\infty} d_k(x, y) = \delta(x, y), \quad \forall x, y \in M.$$

If P is strong Feller, then Proposition 5.23 implies that P^2 is ultra Feller. Therefore

$$0 = \inf \left\{ \sup_{y \in U} \|\delta_x P^2 - \delta_y P^2\|_\delta \ : \ U \text{ open}, x \in U \right\}. \tag{5.4}$$

Since $(d_k)_{k \geq 1}$ is nondecreasing and converges pointwise to δ, the sequence of functions $f_k(y) := \|\delta_x P^2 - \delta_y P^2\|_{d_k}$ is nondecreasing and dominated by $f(y) := \|\delta_x P^2 - \delta_y P^2\|_\delta$. Thus, for every open neighborhood U of x,

$$\limsup_{k \to \infty} \sup_{y \in U} f_k(y) \leq \sup_{y \in U} \lim_{k \to \infty} f_k(y) \leq \sup_{y \in U} f(y).$$

Together with (5.4) and $n_k := 2$ for all $k \geq 1$, this yields

$$\inf \left\{ \limsup_{k \to \infty} \sup_{y \in U} \|\delta_x P^{n_k} - \delta_y P^{n_k}\|_{d_k} \ : \ U \text{ open}, x \in U \right\} = 0.$$

$\square$

Remark 5.27 If P is a Markov kernel on a separable metric space such that P^n is strong Feller for some $n \in \mathbb{N}^*$, then P is asymptotically strong Feller. This follows if one replaces P^2 in the proof of Proposition 5.26 with P^{2n}.

The following exercise shows that the converse of Proposition 5.26 does not hold, i.e., there are Markov kernels which are asymptotically strong Feller but not strong Feller.

Exercise 5.28 Consider the mapping

$$F : \mathbb{R}^2 \to \mathbb{R}^2, \ (x_1, x_2) \mapsto (x_2, x_1).$$

For $(x, \theta) \in \mathbb{R}^2 \times \mathbb{R}$, set $F_\theta(x) := F(x) + \theta e_1$, where $e_1 := (1, 0)^\top$ (cf. Exercise 6.11 (ii) in Sect. 6.2). Let m be a probability measure on $(\mathbb{R}, \mathcal{B}(\mathbb{R}))$ that is absolutely continuous with respect to Lebesgue measure.

(i) Show that the Markov kernel P corresponding to the random dynamical system (F, m) is not strong Feller. *Hint:* Consider for instance the function $f(x_1, x_2) := \mathbf{1}_{x_2 \geq 0}$.

(ii) Use the result from Exercise 6.10 in Sect. 6.2 to show that P^2 is strong Feller, and conclude that P is asymptotically strong Feller.

5.3.2 A Sufficient Condition for the Asymptotic Strong Feller Property

Throughout Sect. 5.3.2, let H be a separable real Hilbert space with norm $\| \cdot \|$, and let $\|f\|_\infty := \sup_{x \in H} |f(x)|$ for $f \in B(H)$, the set of real-valued bounded Borel-measurable functions on H.

Definition 5.29 A function $f : H \to \mathbb{R}$ is called Fréchet differentiable at a point $x \in H$ if there exists a bounded linear operator $A : H \to \mathbb{R}$ such that

$$\lim_{\|h\| \to 0} \frac{|f(x+h) - f(x) - Ah|}{\|h\|} = 0.$$

The operator A is uniquely defined by the above condition, and it is called the Fréchet derivative of f at the point x.

Let $F(H)$ denote the space of bounded functions $f : H \to \mathbb{R}$ that are Fréchet differentiable and whose Fréchet derivative ∇f satisfies the following conditions:

(i)

$$\|\nabla f\|_\infty := \sup_{x \in H} \sup_{h \in H : \|h\| \leq 1} |\nabla f(x)h| < \infty;$$

(ii) The mapping $x \mapsto \nabla f(x)h$ is continuous for every $h \in H$.

The following statement is a special case of Proposition 3.12 in [34].

Theorem 5.30 (Hairer, Mattingly) *Let P be a Markov kernel on $(H, \mathcal{B}(H))$. Assume that there exist constants $\alpha \in (0,1)$ and $C > 0$ such that for every $f \in F(H)$, one has $Pf \in F(H)$ and*

$$\|\nabla Pf\|_\infty \leq C \|f\|_\infty + \alpha \|\nabla f\|_\infty. \tag{5.5}$$

Then P is asymptotically strong Feller.

Proof Consider the sequence of continuous metrics

$$d_k(x,y) := 1 \wedge (\alpha^{-k/2} \|x - y\|), \quad k \in \mathbb{N}^*.$$

Similarly to the sequence of metrics defined in the proof of Proposition 5.26, $(d_k)_{k \geq 1}$ is nondecreasing and converges pointwise to the discrete metric δ. Fix $k \in \mathbb{N}^*$ and $\phi \in \mathrm{Lip}_1(d_k)$. As explained in Remark 5.31 below, there exists a sequence $(\phi_n)_{n \geq 1}$ in $F(H) \cap \mathrm{Lip}_1(d_k)$ such that

$$\lim_{n \to \infty} \phi_n(x) = \phi(x), \quad \forall x \in H.$$

For $x \in H$ and $n \in \mathbb{N}^*$, set

$$\tilde{\phi}(x) := \phi(x) - \sup_{y \in H} \phi(y) \quad \text{and} \quad \tilde{\phi}_n(x) := \phi_n(x) - \sup_{y \in H} \phi_n(y).$$

Notice that $\|\tilde{\phi}\|_\infty = \sup_{y \in H} \phi(y) - \inf_{y \in H} \phi(y) \le 1$ because $\phi \in \text{Lip}_1(d_k)$ and $d_k \le 1$. Similarly, $\|\tilde{\phi}_n\|_\infty \le 1$ for every $n \in \mathbb{N}^*$. Since $\tilde{\phi}_n$ and ϕ_n only differ by a constant, we also have $\tilde{\phi}_n \in F(H) \cap \text{Lip}_1(d_k)$ for every $n \in \mathbb{N}^*$. It is then not hard to see that

$$\|\nabla \tilde{\phi}_n\|_\infty \le \alpha^{-k/2}, \quad \forall n \in \mathbb{N}^*.$$

Now, fix $x, y \in H$ and define

$$\gamma(s) := (1 - s)y + sx, \quad s \in [0, 1].$$

By assumption, $P^k \tilde{\phi}_n \in F(H)$ for every $n \in \mathbb{N}^*$. By the chain rule for the Fréchet derivative, the function $P^k \tilde{\phi}_n \circ \gamma$ is differentiable with

$$(P^k \tilde{\phi}_n \circ \gamma)'(s) = \nabla P^k \tilde{\phi}_n(\gamma(s))(x - y), \quad \forall s \in (0, 1).$$

Since the expression on the right-hand side is continuous in s, one obtains with the fundamental theorem of calculus

$$P^k \phi_n(x) - P^k \phi_n(y) = P^k \tilde{\phi}_n(\gamma(1)) - P^k \tilde{\phi}_n(\gamma(0))$$

$$= \int_0^1 \nabla P^k \tilde{\phi}_n(\gamma(s))(x - y)\, ds \le \|x - y\| \|\nabla P^k \tilde{\phi}_n\|_\infty.$$

Iteratively applying the estimate in (5.5), one has

$$\|\nabla P^k \tilde{\phi}_n\|_\infty \le C \sum_{j=0}^{k-1} \alpha^j \|\tilde{\phi}_n\|_\infty + \alpha^k \|\nabla \tilde{\phi}_n\|_\infty.$$

Since $\|\tilde{\phi}_n\|_\infty \le 1$ and $\|\nabla \tilde{\phi}_n\|_\infty \le \alpha^{-k/2}$, this yields

$$P^k \phi_n(x) - P^k \phi_n(y) \le c(C, \alpha) \|x - y\|,$$

where $c(C, \alpha) := \alpha^{1/2} + C/(1 - \alpha)$. Letting $n \to \infty$, we have by bounded convergence

$$P^k \phi(x) - P^k \phi(y) \le c(C, \alpha) \|x - y\|.$$

As this estimate holds for all $\phi \in \text{Lip}_1(d_k)$,

$$\|\delta_x P^k - \delta_y P^k\|_{d_k} \leq c(C, \alpha)\|x - y\|.$$

Now, for $\epsilon > 0$ fixed, let U be the open $\|\cdot\|$-ball of radius ϵ centered at x. For any $z \in U$,

$$\|\delta_x P^k - \delta_z P^k\|_{d_k} \leq c(C, \alpha)\epsilon,$$

hence

$$\inf\left\{\limsup_{k\to\infty} \sup_{z\in U} \|\delta_x P^k - \delta_z P^k\|_{d_k} : U \text{ open}, x \in U\right\} \leq c(C, \alpha)\epsilon.$$

Since ϵ was arbitrarily chosen, P is asymptotically strong Feller. $\square$

Remark 5.31 The proof of Theorem 5.30 uses the following approximation result: For every $\phi \in \text{Lip}_1(d_k)$ there exists a sequence $(\phi_n)_{n\geq 1}$ in $F(H) \cap \text{Lip}_1(d_k)$ that converges pointwise to ϕ. To see this, let $(e_j)_{j\in\mathcal{J}}$ be a complete orthonormal system in H, where either $\mathcal{J} = \mathbb{N}^*$ or $\mathcal{J} = \{1, \dots, N\}$ for some $N \in \mathbb{N}^*$. For $t \geq 0$, define the bounded linear operator

$$A(t) : H \to H, \quad x \mapsto \sum_{j\in\mathcal{J}} e^{-j^2 t}\langle x, e_j\rangle e_j,$$

where $\langle \cdot, \cdot \rangle$ denotes the inner product on H. The collection of operators $(A(t))_{t\geq 0}$ is a C_0-semigroup on H, and $\|A(t)\|_{op} \leq e^{-t}$ for all $t \geq 0$. For $t > 0$, let

$$Q_t : H \to H, \quad x \mapsto \int_0^t A(2s)x\, ds, \tag{5.6}$$

where the integral is to be interpreted as a Bochner integral. It is not hard to see that Q_t is of trace class, so there is a well-defined Gaussian measure μ_t on $(H, \mathcal{B}(H))$ with mean 0 and covariance operator Q_t. For $n \in \mathbb{N}^*$, define

$$\phi_n(x) := \int_H \phi(A(1/n)x + y)\, \mu_t(dy), \quad x \in H.$$

It is not hard to check that $A(t)(H) \subset Q_t^{1/2}(H)$ for every $t > 0$. Then, by Theorem 2.1 in [56], ϕ_n has Fréchet derivatives of any order, and all derivatives and the function itself are bounded. In particular, $\phi_n \in F(H)$ for all $n \in \mathbb{N}^*$.

For $n \in \mathbb{N}^*$ and $x, y \in H$, one has

$$|\phi_n(x) - \phi_n(y)| \leq \int_H |\phi(A(1/n)x + z) - \phi(A(1/n)y + z)| \, \mu_t(dz)$$

$$\leq \int_H d_k(A(1/n)x + z, A(1/n)y + z) \, \mu_t(dz) \leq d_k(x, y).$$

Finally, the pointwise convergence of $(\phi_n)_{n \geq 1}$ to ϕ follows from Proposition 6.2 in [14].

5.3.3   Unique Ergodicity of Asymptotic Strong Feller Chains

The following theorem, first shown in [34], provides an important justification for introducing the asymptotic strong Feller property. It can be seen as a strengthening of Proposition 5.18 (i) for Polish spaces.

Theorem 5.32 (Hairer, Mattingly)   *Let (M, d^*) be a Polish space, i.e., a complete and separable metric space, and let P be a Markov kernel on $(M, \mathcal{B}(M))$. Let μ, ν be ergodic measures with respect to P. If P is asymptotically strong Feller at a point $x \in \mathsf{supp}(\mu) \cap \mathsf{supp}(\nu)$, then $\mu = \nu$. In particular, if P is asymptotically strong Feller, then two distinct ergodic measures have disjoint support.*

The proof of Theorem 5.32 requires several tools we yet need to introduce. We therefore postpone it to the end of this subsection. Let (X, e) be an arbitrary metric space and let $\mu, \nu \in \mathcal{P}(X)$. A *coupling* of μ and ν is a probability measure Γ on $(X^2, \mathcal{B}(X) \otimes \mathcal{B}(X))$ such that

$$\Gamma(A \times X) = \mu(A), \quad \Gamma(X \times A) = \nu(A), \quad \forall A \in \mathcal{B}(X).$$

We denote by $\mathcal{C}(\mu, \nu)$ the set of couplings of μ and ν.

Exercise 5.33   Assume in addition that X is separable and let $\mathcal{P}(X^2)$ be the set of Borel probability measures on X^2, endowed with the topology of weak convergence. Show that for every $\mu, \nu \in \mathcal{P}(X)$, $\mathcal{C}(\mu, \nu)$ is a closed subset of $\mathcal{P}(X^2)$.

The following exercise explores the concept of lower semicontinuity. Given a metric space (X, e), a function $f : X \to \mathbb{R}$ is called lower semicontinuous if $f(x_0) \leq \liminf_{x \to x_0} f(x)$ for every $x_0 \in X$.

Exercise 5.34   Let $f : X \to [0, \infty)$ be a function.

(i) Show that

$$\tilde{f}(x) := \inf\{f(y) + e(x, y) : y \in X\}$$

defines a continuous function on X.

(ii) Show that f is lower semicontinuous if and only if there exists a nondecreasing sequence $(f_n)_{n \geq 1}$ of continuous functions from X to $[0, \infty)$ that converges pointwise to f. *Hint:* Consider the functions $f_n(x) := \inf\{f(y) + ne(x, y) : y \in X\}$, $n \in \mathbb{N}^*$, and use part (i).

The following statement, cited without proof here, can be found in [68] (see Particular Case 5.16 of Theorem 5.10 for the formula and Theorem 4.1 for existence of a minimizing coupling). It is an instance of the famous Kantorovich–Rubinstein duality theorem. The term *duality* refers to the asserted equivalence of a maximization and a minimization problem.

Theorem 5.35 *Let (M, d^*) be a Polish space and let d be a bounded metric on M that is lower semicontinuous as a function from the product metric space $(M \times M, d^* \star d^*)$ to $[0, \infty)$. Then, for every $\mu, \nu \in \mathcal{P}(M)$, we have*

$$\|\mu - \nu\|_d = \inf_{\Gamma \in \mathcal{C}(\mu, \nu)} \int_{M^2} d(x, y)\, \Gamma(dx, dy)$$

and the infimum on the right-hand side is attained.

Remark 5.36 Let (M, d^*) be a Polish space. The *Wasserstein distance* of order 1 between $\mu, \nu \in \mathcal{P}(M)$ is defined as

$$W_1(\mu, \nu) := \inf_{\Gamma \in \mathcal{C}(\mu, \nu)} \int_{M^2} d^*(x, y)\, \Gamma(dx, dy).$$

In light of Theorem 5.35, if d^* is bounded, then

$$W_1(\mu, \nu) = \|\mu - \nu\|_{d^*}, \quad \forall \mu, \nu \in \mathcal{P}(M).$$

Exercise 5.20 shows that in this case, W_1 is a bounded metric on $\mathcal{P}(M)$. One can show that the metric space $(\mathcal{P}(M), W_1)$ is Polish as well (see, e.g., Theorem 6.18 in [68]).

Lemma 5.37 *Let (M, d^*) be a Polish space, let $(d_n)_{n \geq 1}$ be a nondecreasing sequence of continuous metrics on M, and let d be a bounded metric on M such that*

$$\lim_{n \to \infty} d_n(x, y) = d(x, y), \quad \forall x, y \in M.$$

Then, for every $\mu, \nu \in \mathcal{P}(M)$, we have

$$\lim_{n \to \infty} \|\mu - \nu\|_{d_n} = \|\mu - \nu\|_d.$$

Proof Let $\mu, \nu \in \mathcal{P}(M)$. Since $(d_n)_{n\geq 1}$ is nondecreasing and since d is bounded, we have

$$\|\mu - \nu\|_{d_n} \leq \|\mu - \nu\|_{d_{n+1}} \leq \|\mu - \nu\|_d < \infty, \quad \forall n \in \mathbb{N}^*.$$

Therefore,

$$l := \lim_{n\to\infty} \|\mu - \nu\|_{d_n}$$

exists and is less than or equal to $\|\mu - \nu\|_d$. By Theorem 5.35, there are couplings $(\Gamma_n)_{n\geq 1}$ of μ and ν such that

$$\|\mu - \nu\|_{d_n} = \int_{M^2} d_n(x, y) \, \Gamma_n(dx, dy), \quad \forall n \in \mathbb{N}^*.$$

Since μ and ν are Borel probability measures on a Polish space, they are tight by Prohorov's theorem 4.13, i.e., for every $\varepsilon > 0$ there is a compact set $K \subset M$ such that $\mu(K), \nu(K) > 1 - \varepsilon$. Hence, by Exercise 5.38 below, the family of couplings $(\Gamma_n)_{n\geq 1}$ is tight as well. Again by Prohorov's theorem, $(\Gamma_n)_{n\geq 1}$ admits a subsequence that converges weakly to a probability measure $\Gamma_\infty \in \mathcal{P}(M^2)$. And by Exercise 5.33, $\Gamma_\infty \in \mathcal{C}(\mu, \nu)$. For simplicity, we denote the convergent subsequence again by $(\Gamma_n)_{n\geq 1}$. For $n \leq m$, we have

$$\int_{M^2} d_n(x, y) \, \Gamma_m(dx, dy) \leq \int_{M^2} d_m(x, y) \, \Gamma_m(dx, dy) = \|\mu - \nu\|_{d_m} \leq l.$$

Since each d_n is continuous and bounded, and since Γ_m converges weakly to Γ_∞, we have

$$\lim_{m\to\infty} \int_{M^2} d_n(x, y) \, \Gamma_m(dx, dy) = \int_{M^2} d_n(x, y) \, \Gamma_\infty(dx, dy).$$

Thus,

$$\int_{M^2} d_n(x, y) \, \Gamma_\infty(dx, dy) \leq l.$$

By monotone convergence,

$$l \geq \int_{M^2} d(x, y) \, \Gamma_\infty(dx, dy) \geq \inf_{\Gamma \in \mathcal{C}(\mu,\nu)} \int_{M^2} d(x, y) \, \Gamma(dx, dy). \tag{5.7}$$

Since d is the pointwise limit of a nondecreasing sequence of continuous functions, Exercise 5.34 implies that d is lower semicontinuous. Hence, by virtue of Theorem 5.35, the expression on the right-hand side of (5.7) equals $\|\mu - \nu\|_d$. We have thus shown that $l \geq \|\mu - \nu\|_d$, and together with $l \leq \|\mu - \nu\|_d$ one obtains $\lim_{n \to \infty} \|\mu - \nu\|_{d_n} = \|\mu - \nu\|_d$. $\square$

Exercise 5.38 Let (X, e) be a metric space and let $\mu, \nu \in \mathcal{P}(X)$ be tight. Show that $\mathcal{C}(\mu, \nu) \subset \mathcal{P}(X^2)$ is a tight family of probability measures.

Lemma 5.39 *Let (M, d^*) be a separable metric space, let P be a Markov kernel on $(M, \mathcal{B}(M))$, and let d be a metric on M that is bounded by 1. Assume further that there are $\epsilon > 0$ and $U \in \mathcal{B}(M)$ such that*

$$\sup_{x,y \in U} \|\delta_x P - \delta_y P\|_d \leq \epsilon.$$

Let $\mu, \nu \in \mathcal{P}(M)$ and set $\alpha := \mu(U) \wedge \nu(U)$. Then

$$\|\mu P - \nu P\|_d \leq 1 - \alpha(1 - \epsilon).$$

Proof Since d is bounded by 1, we have $\|\mu P - \nu P\|_d \leq 1$, so the assertion holds if $\alpha = 0$. If $\alpha > 0$, define for $A \in \mathcal{B}(M)$ the Borel probability measures

$$\mu^U(A) := \frac{\mu(A \cap U)}{\mu(U)}, \qquad\qquad \nu^U(A) := \frac{\nu(A \cap U)}{\nu(U)},$$

$$\bar{\mu}(A) := \frac{\mu(A) - \alpha \mu^U(A)}{1 - \alpha}, \qquad\qquad \bar{\nu}(A) := \frac{\nu(A) - \alpha \nu^U(A)}{1 - \alpha},$$

and observe that

$$\mu = (1 - \alpha)\bar{\mu} + \alpha \mu^U,$$

$$\nu = (1 - \alpha)\bar{\nu} + \alpha \nu^U.$$

Let $\phi \in \mathrm{Lip}_1(d)$. Exercise 5.40 below and the fact that $\mu^U(U^c) = \nu^U(U^c) = 0$ yield

$$(\mu^U P)\phi - (\nu^U P)\phi = \int_{U^2} \left((\delta_x P)\phi - (\delta_y P)\phi\right) \, \mu^U(dx)\, \nu^U(dy)$$

$$\leq \int_{U^2} \|\delta_x P - \delta_y P\|_d \, \mu^U(dx)\, \nu^U(dy) \leq \epsilon.$$

Taking the supremum over $\mathrm{Lip}_1(d)$ gives

$$\|\mu^U P - \nu^U P\|_d \leq \epsilon.$$

The triangle inequality for $\| \cdot \|_d$ then implies

$$\|\mu P - \nu P\|_d \leq (1 - \alpha)\|\bar{\mu} P - \bar{\nu} P\|_d + \alpha \|\mu^U P - \nu^U P\|_d \leq 1 - \alpha + \alpha\epsilon.$$

$\square$

Exercise 5.40 Let (M, d^*) be a separable metric space, let P be a Markov kernel on $(M, \mathcal{B}(M))$, and let d be a bounded metric on M. Show that for every $\mu, \nu \in \mathcal{P}(M)$, $\Gamma \in \mathcal{C}(\mu, \nu)$, and $\phi \in \mathrm{Lip}_1(d)$, one has

$$(\mu P)\phi - (\nu P)\phi = \int_{M^2} \left((\delta_x P)\phi - (\delta_y P)\phi\right) \Gamma(dx, dy).$$

We are now ready to prove Theorem 5.32.

Proof *(Theorem 5.32)* Let $x \in \mathrm{supp}(\mu) \cap \mathrm{supp}(\nu)$ such that P is asymptotically strong Feller at x. Then there exist a nondecreasing sequence $(n_k)_{k \geq 1}$ of positive integers as well as a nondecreasing sequence $(d_k)_{k \geq 1}$ of continuous metrics on M such that $\lim_{k \to \infty} d_k(y, z) = \delta(y, z)$, $y, z \in M$, and

$$\inf_{\substack{x \in U \subset M, \\ U \, open}} \limsup_{k \to \infty} \sup_{y \in U} \|\delta_x P^{n_k} - \delta_y P^{n_k}\|_{d_k} = 0.$$

Let U be an open neighborhood of x and let $K \in \mathbb{N}$ such that

$$\sup_{y \in U} \|\delta_x P^{n_k} - \delta_y P^{n_k}\|_{d_k} < \frac{1}{4}, \quad \forall k \geq K.$$

Since $\| \cdot \|_d$ satisfies the triangle inequality for every metric d on (M, d^*), we have

$$\sup_{y,z \in U} \|\delta_y P^{n_k} - \delta_z P^{n_k}\|_{d_k} < \frac{1}{2}, \quad \forall k \geq K.$$

Set $\alpha := \mu(U) \wedge \nu(U)$. Lemma 5.39 implies

$$\|\mu P^{n_k} - \nu P^{n_k}\|_{d_k} \leq 1 - \frac{\alpha}{2}, \quad \forall k \geq K.$$

Since μ and ν are invariant probability measures,

$$\|\mu - \nu\|_{d_k} \leq 1 - \frac{\alpha}{2}, \quad \forall k \geq K.$$

As

$$\lim_{k \to \infty} \|\mu - \nu\|_{d_k} = \|\mu - \nu\|_\delta$$

by Lemma 5.37, it follows that

$$\|\mu - \nu\|_\delta \leq 1 - \frac{\alpha}{2}.$$

Since $x \in \mathsf{supp}(\mu) \cap \mathsf{supp}(\nu)$, we have $\alpha > 0$, so $\|\mu - \nu\|_\delta < 1$. In particular, for every $A \in \mathcal{B}(M)$,

$$2 > |\mu(\mathbf{1}_A - \mathbf{1}_{A^c}) - \nu(\mathbf{1}_A - \mathbf{1}_{A^c})| = 2|\mu(A) - \nu(A)|$$

in view of Remark 5.21. This implies that μ and ν are not mutually singular. Since μ and ν are ergodic, it follows from Proposition 4.29 (ii) that $\mu = \nu$.   □

For a Markov kernel P, let $\mathsf{Erg}(P)$ denote the set of P-ergodic measures. From the proof of Theorem 5.32, one obtains the following corollary.

Corollary 5.41 *Let M be a Polish space and let P be asymptotically strong Feller at a point $x \in M$. Then there exist a neighborhood U of x and an ergodic measure ν such that $\pi(U) = 0$ for every $\pi \in \mathsf{Erg}(P) \setminus \{\nu\}$.*

Proof Suppose the statement does not hold. Then for every neighborhood U of x there are at least two distinct $\nu_1, \nu_2 \in \mathsf{Erg}(P)$ such that $\nu_1(U), \nu_2(U) > 0$. As in the proof of Theorem 5.32, one then shows existence of distinct $\nu_1, \nu_2 \in \mathsf{Erg}(P)$ that are not mutually singular, which contradicts Proposition 4.29.   □

In the following proposition, we exploit Theorem 5.32 and its corollary to further elucidate the structure of $\mathsf{Erg}(P)$ under the asymptotic strong Feller property. In particular, we obtain a counterpart of Proposition 5.18 (iii).

Proposition 5.42 *Let M be a Polish space and let P be asymptotically strong Feller.*

  (i) *The set $\mathsf{Erg}(P)$ is countable, and for every P-invariant probability measure μ one has*

$$\mu(\cdot) = \sum_{\nu \in \mathsf{Erg}(P)} \nu(\cdot)\mu(X(\nu)),$$

  where $X(\nu) = \{x \in M : Q(x, \mathsf{supp}\,\nu) = 1\}$ and Q is the Markov kernel from Theorem 4.51;

 (ii) *If P has an invariant probability measure having full support, then*

$$M = \overset{*}{\bigcup_{\nu \in \mathsf{Erg}(P)}} \mathsf{supp}\,\nu,$$

  where the asterisk indicates that $\mathsf{supp}\,\nu_1 \cap \mathsf{supp}\,\nu_2 = \emptyset$ for distinct $\nu_1, \nu_2 \in \mathsf{Erg}(P)$;

(iii) *If P has an invariant probability measure μ having full support and if M is connected, then* $\mathsf{Erg}(P)$ *is either countably infinite or* $\mathsf{Erg}(P) = \{\mu\}$;

(iv) *Suppose that P has an invariant probability measure μ having full support. Assume in addition that for every $\varepsilon > 0$ there exists a connected compact set $K \subset M$ such that $\mu(K) > 1 - \varepsilon$. Then* $\mathsf{Erg}(P) = \{\mu\}$.

Remark 5.43 The condition from part (iv) that for every $\varepsilon > 0$ there is a connected compact set $K \subset M$ with $\mu(K) > 1 - \varepsilon$ clearly holds if M is connected and compact. But it also holds, for instance, if M is a separable Banach space or, more generally, a separable Fréchet space. By Fréchet space we mean a locally convex topological vector space whose topology is induced by a complete metric d that satisfies $d(x + z, y + z) = d(x, y)$ for every $x, y, z \in M$. Indeed, since Borel probability measures on a Polish space are tight, for every $\varepsilon > 0$ there is a compact set $\tilde{K} \subset M$ such that $\mu(\tilde{K}) > 1 - \varepsilon$. Let K be the closure of the convex hull of $\tilde{K}$. Then $\mu(K) > 1 - \varepsilon$ and K is convex, hence connected. By Theorem 3.20 (c) in [61], K is also compact as the closure of the convex hull of a compact set in a Fréchet space.

The following lemma is used in the proof of Proposition 5.42.

Lemma 5.44 *Let M be a Polish space and let P asymptotically strong Feller at a point $x \in M$. If there is an invariant probability measure μ such that $x \in \mathsf{supp}\,\mu$, then $x \in \mathsf{supp}\,\nu$ for some $\nu \in \mathsf{Erg}(P)$.*

Proof By Corollary 5.41, there are a neighborhood U_x of x and $\nu \in \mathsf{Erg}(P)$ such that $\pi(U_x) = 0$ for every $\pi \in \mathsf{Erg}(P) \setminus \{\nu\}$. To see that $x \in \mathsf{supp}\,\nu$, fix a neighborhood U of x. Then $U \cap U_x$ is also a neighborhood of x. Since $x \in \mathsf{supp}\,\mu$, one has $\mu(U \cap U_x) > 0$. By the ergodic decomposition theorem 4.51, there is a Markov kernel Q such that $Q(y, \cdot) \in \mathsf{Erg}(P)$ for μ-almost every $y \in M$, and

$$0 < \mu(U \cap U_x) = \int_M Q(y, U \cap U_x)\,\mu(dy).$$

Hence, there is $y \in M$ such that $Q(y, \cdot) \in \mathsf{Erg}(P)$ and $Q(y, U \cap U_x) > 0$. It follows that $Q(y, U_x) > 0$, so $Q(y, \cdot) = \nu$. Consequently,

$$\nu(U) = Q(y, U) \geq Q(y, U \cap U_x) > 0.$$

$\square$

Proof *(Proposition 5.42)*

(i) Since M is separable, so is its subset $S := \bigcup_{\nu \in \mathsf{Erg}(P)} \mathsf{supp}\,\nu$ (see Exercise 4.45 (i)). Let D be countable and dense in S. By Theorem 5.32, the supports of distinct P-ergodic measures are disjoint. To show that $\mathsf{Erg}(P)$ is countable, it is enough to prove that for every $\nu \in \mathsf{Erg}(P)$ there is $x \in D$ with $x \notin \bigcup_{\pi \in \mathsf{Erg}(P) \setminus \{\nu\}} \mathsf{supp}\,\pi$. Let $\nu \in \mathsf{Erg}(P)$ and let $y \in \mathsf{supp}\,\nu$. By

Corollary 5.41, there is an open neighborhood U of y such that $\pi(U) = 0$ for every $\pi \in \mathsf{Erg}(P) \setminus \{v\}$. Since $y \in S$ and since D is dense in S, there is a point $x \in D \cap U$. As U is a neighborhood of x, one has $x \notin \mathsf{supp}\,\pi$ for every $\pi \in \mathsf{Erg}(P) \setminus \{v\}$.

By Theorem 5.32, $X(v) \cap X(\pi) = \emptyset$ whenever $v, \pi \in \mathsf{Erg}(P)$ are distinct. Let

$$\mathcal{X} = \{x \in M :\ Q(x, \cdot) \in \mathsf{Erg}(P)\}.$$

Theorem 4.51 implies that there is $B \in \mathcal{B}(M)$ such that $B \subset \mathcal{X}$ and $\mu(B) = 1$. Since $\mathcal{X} \subset \bigcup_{v \in \mathsf{Erg}(P)} X(v)$, one has

$$B \subset \bigcup_{v \in \mathsf{Erg}(P)} X(v)$$

and hence

$$1 = \mu(B) = \mu\left(B \cap \bigcup_{v \in \mathsf{Erg}(P)} X(v)\right) = \mu\left(\bigcup_{v \in \mathsf{Erg}(P)} (B \cap X(v))\right).$$

With Theorem 4.51, this yields for every $A \in \mathcal{B}(M)$

$$\mu(A) = \int_M Q(x, A)\,\mu(dx) = \sum_{v \in \mathsf{Erg}(P)} \int_{B \cap X(v)} Q(x, A)\,\mu(dx). \qquad (5.8)$$

For $v \in \mathsf{Erg}(P)$, let $x \in B \cap X(v)$. Since $x \in B \subset \mathcal{X}$, we have $Q(x, \cdot) \in \mathsf{Erg}(P)$. And since $x \in X(v)$, we have $Q(x, \mathsf{supp}\,v) = 1$. Then, by Theorem 5.32, $Q(x, \mathsf{supp}\,\pi) = 0$ for every $\pi \in \mathsf{Erg}(P) \setminus \{v\}$. In particular, $Q(x, \cdot) = v$. Thus, the expression on the right-hand side of (5.8) equals

$$\sum_{v \in \mathsf{Erg}(P)} \int_{B \cap X(v)} v(A)\,\mu(dx) = \sum_{v \in \mathsf{Erg}(P)} v(A)\mu(B \cap X(v)) = \sum_{v \in \mathsf{Erg}(P)} v(A)\mu(X(v)).$$

(ii) This follows immediately from Lemma 5.44 and Theorem 5.32.
(iii) In light of part (i) and the ergodic decomposition theorem, all we need to show is that if $\mathsf{Erg}(P)$ is finite, then it has cardinality 1. Suppose that $\mathsf{Erg}(P) = \{v_1, \ldots, v_n\}$, where n is a positive integer and $v_1, \ldots, v_n$ are pairwise distinct. By part (ii), M is a finite and disjoint union of nonempty closed sets. As M is connected, this is only possible if $n = 1$.
(iv) The following claim will be proved later on:

Claim 1 If $K \subset M$ is connected and compact, then there is $v \in \mathsf{Erg}(P)$ such that $\pi(K) = 0$ for every $\pi \in \mathsf{Erg}(P) \setminus \{v\}$.

We use Claim 1 now to show

Claim 2 There are $S \in \mathcal{B}(M)$ and $\nu \in \mathsf{Erg}(P)$ such that $\mu(S) = 1$ and $\pi(S) = 0$ for every $\pi \in \mathsf{Erg}(P) \setminus \{\nu\}$.

By assumption, for every integer $n \geq 2$ there is a connected compact set $K_n \subset M$ such that $\mu(K_n) > 1 - \frac{1}{n}$. Set

$$S := \bigcup_{n \geq 2} K_n.$$

Then, for every $m \geq 2$,

$$\mu(S) \geq \mu(K_m) > 1 - \tfrac{1}{m},$$

which implies $\mu(S) = 1$.

Claim 1 implies that for every $n \geq 2$ there is $\nu_n \in \mathsf{Erg}(P)$ such that $\pi(K_n) = 0$ for every $\pi \in \mathsf{Erg}(P) \setminus \{\nu_n\}$. Set $\nu := \nu_2$. To show that $\pi(S) = 0$ for every $\pi \in \mathsf{Erg}(P) \setminus \{\nu\}$, it is then sufficient to prove $\nu_n = \nu$ for every $n > 2$. Suppose this is not the case. Then there is $n > 2$ such that $\nu_n \neq \nu$. Since $\pi(K_n) = 0$ for every $\pi \in \mathsf{Erg}(P) \setminus \{\nu_n\}$ and $\pi(K_2) = 0$ for every $\pi \in \mathsf{Erg}(P) \setminus \{\nu\}$, we have in particular $\pi(K_n \cap K_2) = 0$ for every $\pi \in \mathsf{Erg}(P)$. On the other hand,

$$\mu(K_n \cap K_2) = \mu(K_n) + \mu(K_2) - \mu(K_n \cup K_2) > 2 - \frac{1}{n} - \frac{1}{2} - 1 > 0.$$

Hence, by part (i),

$$0 < \mu(K_n \cap K_2) = \sum_{\pi \in \mathsf{Erg}(P)} \pi(K_n \cap K_2)\mu(X(\pi)) = 0,$$

a contradiction. This completes the proof of Claim 2.

Let S and ν be as stipulated in Claim 2. By the formula in part (i),

$$1 = \mu(S) = \sum_{\pi \in \mathsf{Erg}(P)} \pi(S)\mu(X(\pi)) = \nu(S)\mu(X(\nu)).$$

In particular, $\mu(X(\nu)) = 1$. Since $X(\pi) \cap X(\psi) = \emptyset$ for distinct $\pi, \psi \in \mathsf{Erg}(P)$, this yields $\mu = \nu$, so $\mu \in \mathsf{Erg}(P)$. As $\mathsf{supp}\, \mu = M$ and as distinct P-ergodic measures have disjoint supports, one obtains $\mathsf{Erg}(P) = \{\mu\}$.

Finally we need to prove Claim 1. By Corollary 5.41, we can associate every $y \in K$ with an open neighborhood U_y of y and $\nu_y \in \mathsf{Erg}(P)$ such that $\pi(U_y) = 0$ for every $\pi \in \mathsf{Erg}(P) \setminus \{\nu_y\}$. Since K is compact, there are finitely many

$y_1, \ldots, y_n \in K$ such that

$$K \subset \bigcup_{k=1}^{n} U_{y_k}. \tag{5.9}$$

To simplify notation, we write ν_i instead of ν_{y_i} from now on. Let $\pi \in \mathsf{Erg}(P) \setminus \{\nu_1, \ldots, \nu_n\}$. Then

$$\pi(K) \leq \sum_{k=1}^{n} \pi(U_{y_k}) = 0.$$

To prove Claim 1, it remains to show that $\nu_1 = \ldots = \nu_n$. For $1 \leq i \leq n$, let $F_i := \mathsf{supp}\, \nu_i \cap K$. As the intersection of two closed sets, each set F_i is closed. Besides, F_i is nonempty: Clearly $y_i \in K$, and since $\pi(U_{y_i}) = 0$ for every $\pi \in \mathsf{Erg}(P) \setminus \{\nu_i\}$, part (ii) yields $y_i \in \mathsf{supp}\, \nu_i$. Together with (5.9), part (ii) also implies $K = \bigcup_{i=1}^{n} F_i$. Moreover, $F_i \cap F_j = \emptyset$ if $\nu_i \neq \nu_j$. Connectedness of K then yields $\nu_1 = \ldots = \nu_n$. □

Notes

The notion of ξ-irreducibility introduced at the beginning of Sect. 5.1 is called φ-irreducibility in [49]. For the resolvent kernel R_a, Meyn and Tweedie [49] use the notation K_{a_ε}, where $\varepsilon \in (0, 1)$ corresponds to our parameter a. Section 4.5 of [49] contains additional information, some of it bibliographic, on the use of irreducibility in the study of Markov chains.

The original definition of the asymptotic strong Feller property in [34] is for Markov semigroups (P_t), where $t \geq 0$ is a continuous-time parameter. Translating the definition as well as the results of Hairer and Mattingly to the discrete-time setting is straightforward. Furthermore, the nondecreasing sequence $(d_k)_{k \geq 1}$ converging to δ is allowed to consist of pseudometrics in [34], i.e., the distances between distinct points need not be strictly positive.

Most of the material in Sect. 5.3 is taken from [34], sometimes with small adaptations (in particular, Proposition 5.26, Theorem 5.30, and Theorem 5.32 along with their proofs, including Lemmas 5.37 and 5.39). As far as we know, the statements in Proposition 5.42 have not been published elsewhere.

In the Kantorovich–Rubinstein duality theorem 5.35, the boundedness assumption on the metric d can be relaxed, see [68]. If d^* is unbounded, the Wasserstein distance W_1 defined in Remark 5.36 is still a metric on

$$\mathcal{P}_1(M) := \left\{ \mu \in \mathcal{P}(M) : \int_M d^*(x, y)\, \mu(dy) < \infty \right\},$$

the so-called *Wasserstein space* of order 1. Notice that the choice of x in the definition of $\mathcal{P}_1(M)$ is arbitrary.

Chapter 6
Petite Sets and Doeblin Points

Often, the ξ-irreducibility property, as defined in Chap. 5, can be deduced from the existence of an *accessible point* satisfying a *local Doeblin condition*. These conditions turn out to be very useful tools when dealing with specific models such as random dynamical systems, processes obtained by random switching between deterministic differential equations, or stochastic differential equations. For these models, the accessibility condition can be rewritten as a deterministic control problem and the local Doeblin conditions can be deduced from more "computable" conditions such as—for the last two models—certain Hörmander-type conditions. This chapter develops these ideas in detail.

6.1 Petite Sets, Small Sets, Doeblin Points

We call a measurable set C a *petite set* if there exist $a \in (0, 1)$ and some nonzero Borel measure ξ on M such that

$$R_a(x, A) \geq \xi(A)$$

for all $x \in C$ and $A \in \mathcal{B}(M)$. We call the set C a *small set* if there is a nonzero Borel measure ξ on M such that

$$P(x, A) \geq \xi(A)$$

for all $x \in C$ and $A \in \mathcal{B}(M)$. Clearly, every small set is petite.

M. Benaïm, T. Hurth, *Markov Chains on Metric Spaces*, Universitext,
https://doi.org/10.1007/978-3-031-11822-7_6

Remark 6.1 In the terminology of Meyn and Tweedie [49] (Chapter 5), a ν_α-*petite set* for a probability measure α on $\mathbb{N}$ is a set $C \in \mathcal{B}(M)$ such that

$$\sum_{n=0}^{\infty} \alpha(n) P^n(x, A) \geq \nu_\alpha(A), \quad \forall x \in C, \ A \in \mathcal{B}(M),$$

where ν_α is some nonzero Borel measure on M. A ν_m-*small set* for $m \in \mathbb{N}^*$ is a set $C \in \mathcal{B}(M)$ such that

$$P^m(x, A) \geq \nu_m(A), \quad \forall x \in C, \ A \in \mathcal{B}(M),$$

where ν_m is a nonzero Borel measure on M. With these definitions, the class of petite sets defined above is equal to the class of sets that are ν_{Δ_a}-petite for some $a \in (0, 1)$, where

$$\Delta_a(k) := a^k(1 - a), \ k \in \mathbb{N}.$$

Our notion of a small set corresponds to the notion of a ν_1-small set.

We call a point $x^* \in M$ a *weak Doeblin point* (respectively a *Doeblin point*) if x^* has a neighborhood that is a petite set (respectively a small set).

The importance and usefulness of these notions will be highlighted in Chaps. 7 and 8. Here we mainly focus on weak Doeblin points. The following proposition extends Example 5.3. It provides a powerful tool to ensure unique ergodicity.

Theorem 6.2 *Assume that there exists an accessible weak Doeblin point for P. Then P is ξ-irreducible. In particular, by Theorem 5.5, it has at most one invariant probability measure.*

Proof By assumption, there exists an open set C and a nontrivial measure ξ such that $C \cap \Gamma \neq \emptyset$ and $R_a(x, \cdot) \geq \xi(\cdot)$ for all $x \in C$. Let $p_k = \sum_{i=0}^{k}(1-a)^2 a^i a^{k-i} = (k+1)(1-a)^2 a^k$. Then, for all A measurable and $x \in M$,

$$\sum_{k \geq 0} p_k P^k(x, A) = R_a^2(x, A) = \int R_a(x, dy) R_a(y, A) \geq R_a(x, C) \xi(A).$$

By accessibility, $R_a(x, C) > 0$. $\square$

6.1.1 Continuous Time: Doeblin Points for Markov Processes

Let $\{P_t\}_{t \geq 0}$ be a continuous-time Markov semigroup. Recall (see Sect. 5.2.1) that a point $p \in M$ is called accessible for $\{P_t\}_{t \geq 0}$ provided that it is accessible for the 1-resolvent G, or equivalently, $G(x, U) > 0$ for every $x \in M$ and for every neighborhood U of p. The following proposition is a useful tool whose proof is based on ideas borrowed from [6] and [10].

Proposition 6.3 *Let $\{P_t\}_{t\geq0}$ be a continuous-time weak Feller semigroup. Assume that there exists a point $p \in M$ which is accessible for $\{P_t\}_{t\geq0}$ and which is a Doeblin point for some P_{T_0} with $T_0 > 0$. Then the following statements hold:*

(i) *There exist $q \in M$ (which can be chosen arbitrarily close to p) and $T_1 \geq T_0$ such that for all $T \geq T_1, q$ is an accessible Doeblin point for P_T;*
(ii) *If for some $s > 0$ there exists an invariant probability measure μ for P_s, then μ is the unique invariant probability measure of P_t for all $t > 0$.*

Remark 6.4 Proposition 6.3 is clearly false in discrete time. Let $M = \{0, 1\}$ and $P = \begin{pmatrix} 0 & 1 \\ 1 & 0 \end{pmatrix}$. The point 0 is an accessible Doeblin point (take $\xi = \delta_1$) but is not accessible for P^2.

The proposition also fails to hold if we replace the condition that p is a Doeblin point for some P_{T_0} by the weaker condition that it is a Doeblin point for G. To see this, let $\{P_t\}$ be the semigroup induced by the rotation $x \mapsto (x+t) \mod 1$ on $\mathbb{R}/\mathbb{Z}$ (see Example 4.55). Then every point $p \in M = \mathbb{R}/\mathbb{Z}$ is accessible and a Doeblin point for G, but not accessible for P_α when α is rational.

Proof of Proposition 6.3 By assumption there exists a neighborhood U of p and a nontrivial measure ξ such that for all $x \in U$

$$P_{T_0}(x, \cdot) \geq \xi(\cdot).$$

Lemma 6.5 *There exist $T_0' \geq T_0$, $\varepsilon > 0$, and a measure ζ such that $\zeta(U) > 0$ and*

$$P_t(x, \cdot) \geq \zeta(\cdot)$$

for all $x \in U$ and $T_0' \leq t \leq T_0' + \varepsilon$.

Proof By accessibility, $\xi G(U) = \int_0^\infty e^{-t} \xi P_t(U)\, dt > 0$. Thus, for some $t_0 > 0$, $\xi P_{t_0}(U) > 0$. Set $\xi' = \xi P_{t_0}$. Then $\xi'(U) > 0$ and for all $x \in U$

$$\delta_x P_{T_0+t_0} = \delta_x P_{T_0} P_{t_0} \geq \xi'.$$

By Fatou's lemma, weak Feller continuity and the Portmanteau theorem 4.1,

$$\lim_{s\downarrow0} \xi' P_s(U) \geq \int \liminf_{s\to0} P_s(x, U)\, \xi(dx) \geq \int \mathbf{1}_U(x)\, \xi'(dx) = \xi'(U) > 0.$$

Thus, for some $\delta > 0$ and $\varepsilon > 0$,

$$\xi' P_s(U) \geq \delta$$

for all $0 \leq s \leq \varepsilon$. Set $T_0' = 2(T_0 + t_0)$. Then, for all $x \in U$ and $0 \leq s \leq \varepsilon$,

$$\delta_x P_{T_0'+s} = \delta_x P_{T_0+t_0} P_{T_0+t_0+s} \geq \xi' P_{T_0+t_0+s} \geq \int_U \xi' P_s(dy) P_{T_0+t_0}(y, \cdot) \geq \delta \xi'.$$

This proves the lemma with $\zeta = \delta \xi'$.

We now prove the first part of Proposition 6.3. Set $T_1 = T_0'^2/\varepsilon$. Let $T \geq T_1$. Then T can be written as $T = k(T_0' + s)$ with $0 \leq s \leq \varepsilon$ and $k \in \mathbb{N}^*$. Thus, for all $x \in U$,

$$P_T(x, \cdot) = \delta_x P_T = \delta_x P_{T_0'+s}^k \geq \zeta(U)^{k-1} \zeta.$$

This proves that every point $x \in U$ is a Doeblin point for P_T. Choose now $q \in U \cap \mathrm{supp}(\zeta)$. Let $x \in M$. By accessibility, there exists $t_x > 0$ such that $P_{t_x}(x, U) > 0$. For $k, m \in \mathbb{N}$ sufficiently large, there exists $t \in [T_0', T_0' + \varepsilon]$ such that $t_x + kt = mT$. Then, for every neighborhood V of q,

$$P_T^m(x, V) = P_{t_x+kt}(x, V) \geq P_{t_x}(x, U)\zeta(V)^k > 0.$$

This proves that q is accessible for P_T.

The second assertion follows from Theorem 6.2. Suppose that μ is invariant for some P_s. Then $\nu = \frac{1}{s} \int_0^s \mu P_u \, du$ is invariant for $\{P_t\}$ by Proposition 4.56. Thus, ν is the unique invariant probability measure of P_t for $t \geq T_1$. The same is true for $0 < t \leq T_1$ because $kt \geq T_1$ for some $k \in \mathbb{N}$.   □

6.2   Random Dynamical Systems

Let Θ be a nonempty open subset of $\mathbb{R}^d$ and m a probability measure on $(\Theta, \mathcal{B}(\Theta))$. Let M be a nonempty open subset of $\mathbb{R}^k$ and $F : \Theta \times M \to M$ a C^1-mapping. Recall from Chap. 3 that the pair (F, m) induces a random dynamical system with associated Feller Markov kernel

$$P(x, G) = m(\{\theta \in \Theta : F_\theta(x) \in G\}), \quad (x, G) \in M \times \mathcal{B}(M).$$

For $n \in \mathbb{N}^*$ and $x \in M$, let

$$\varphi_{n,x} : \Theta^n \to M, \quad (\theta_1, \ldots, \theta_n) \mapsto (F_{\theta_n} \circ \ldots \circ F_{\theta_1})(x).$$

The following proposition is essentially Lemma 6.3 in [8].

Proposition 6.6 *Let $x^* \in M$, $n \in \mathbb{N}^*$, and $\theta^* = (\theta_1^*, \ldots, \theta_n^*) \in \Theta^n$ such that the following conditions hold.*

(a) *The Jacobian matrix $D\varphi_{n,x^*}(\theta)|_{\theta=\theta^*}$ has maximal rank (i.e., rank k);*
(b) *There is a neighborhood $V \subset \Theta^n$ of θ^* such that $m^n(\cdot \cap V)$ is absolutely continuous with respect to $\lambda^{nd}(\cdot \cap V)$, where λ^{nd} is the Lebesgue measure on $\mathbb{R}^{nd}$. The corresponding probability density function ρ is such that*

$$c := \inf_{\theta \in V} \rho(\theta) > 0.$$

Under these conditions, x^* is a Doeblin point with respect to the Markov kernel P^n, and in particular a weak Doeblin point with respect to P.

Remark 6.7 Condition (b) above holds true whenever m is absolutely continuous with respect to λ^d with a lower semicontinuous and positive density.

Proof Since $D\varphi_{n,x^*}(\theta)|_{\theta=\theta^*}$ is a $(k \times nd)$-matrix of rank k, we have either $k = nd$ or $k < nd$. To avoid repeating ourselves, we will only prove the slightly more complicated case $k < nd$. The case $k = nd$ can be easily derived by making small modifications to the proof for $k < nd$. Assume without loss of generality that the first k columns of $D\varphi_{n,x^*}(\theta)|_{\theta=\theta^*}$ are linearly independent. We will often write points $\theta \in \Theta^n$ as $\theta = (\theta^{(k)}, \theta^{(nd-k)})$, where $\theta^{(k)} \in \mathbb{R}^k$ is the vector consisting of the first k components of θ, and where $\theta^{(nd-k)}$ is the vector of the remaining $(nd - k)$ components. For $x \in M$, consider the C^1-mapping

$$G_x : \Theta^n \to M \times \mathbb{R}^{nd-k}, \; \theta = (\theta^{(k)}, \theta^{(nd-k)}) \mapsto (\varphi_{n,x}(\theta), \theta^{(nd-k)}).$$

We also define the C^1-mapping

$$H : \Theta^n \times M \to M \times \mathbb{R}^{nd-k} \times M, \; (\theta, x) \mapsto (G_x(\theta), x).$$

Since

$$\det DH(\theta, x)|_{\theta=\theta^*, x=x^*} = \det DG_{x^*}(\theta)|_{\theta=\theta^*}$$
$$= \det D_{\theta^{(k)}} \varphi_{n,x^*}(\theta^{(k)}, (\theta^*)^{(nd-k)})|_{\theta^{(k)}=(\theta^*)^{(k)}} \neq 0,$$

the inverse function theorem implies that there is an open neighborhood W of (θ^*, x^*) such that the restriction of H to W, denoted by H_W, is a C^1-diffeomorphism. By intersecting W with an open subset of $V \times M$ that contains (θ^*, x^*) and calling the resulting set W again, we may assume without loss of generality that $\theta \in V$ for every $(\theta, x) \in W$. The set $H(W)$ is a neighborhood of $H(\theta^*, x^*) = (\varphi_{n,x^*}(\theta^*), (\theta^*)^{(nd-k)}, x^*)$, so there are open neighborhoods Z_0 of $\varphi_{n,x^*}(\theta^*)$, T_0 of $(\theta^*)^{(nd-k)}$, and U_0 of x^* such that $Z_0 \times T_0 \times U_0 \subset H(W)$. Let

$W_0 := H_W^{-1}(Z_0 \times T_0 \times U_0)$. For $x \in U_0$, set

$$V_x := \{\theta \in \Theta^n : (\theta, x) \in W_0\}.$$

It is straightforward to check that for every $x \in U_0$, the restriction of G_x to V_x is a C^1-diffeomorphism that satisfies $G_x(V_x) = Z_0 \times T_0$.

Let $x \in U_0$ and $A \in \mathcal{B}(M)$. We have

$$P^n(x, A) \geq P^n(x, A \cap Z_0) = \int_{\varphi_{n,x}^{-1}(A \cap Z_0)} m^n(d\theta).$$

Since $G_x^{-1}((A \cap Z_0) \times T_0) \subset \varphi_{n,x}^{-1}(A \cap Z_0)$, the expression on the right-hand side is bounded from below by

$$\int_{G_x^{-1}((A \cap Z_0) \times T_0)} m^n(d\theta) \geq \int_{V_x \cap G_x^{-1}((A \cap Z_0) \times T_0)} m^n(d\theta).$$

As $V_x \subset V$, the integral on the right-hand side equals

$$\int_{V_x \cap G_x^{-1}((A \cap Z_0) \times T_0)} \rho(\theta)\, \lambda^{nd}(d\theta) \geq c \int_{V_x \cap G_x^{-1}((A \cap Z_0) \times T_0)} \lambda^{nd}(d\theta).$$

There is no loss of generality in assuming that V and U_0 are each contained in a compact set. Since the mapping $(\theta, x) \mapsto \det DG_x(\theta)$ is continuous, we have

$$\hat{c} := \sup_{\theta \in V, x \in U_0} |\det DG_x(\theta)| < \infty.$$

Hence,

$$P^n(x, A) \geq \frac{c}{\hat{c}} \int_{V_x \cap G_x^{-1}((A \cap Z_0) \times T_0)} |\det DG_x(\theta)|\, \lambda^{nd}(d\theta).$$

Since the restriction of G_x to V_x is a diffeomorphism, the change of variables formula (see for instance Theorem 2.47 in [27]) implies that the expression on the right-hand side equals

$$\frac{c}{\hat{c}} \lambda^{nd-k}(T_0)\lambda^k(A \cap Z_0).$$

The measure $\xi(A) := \frac{c}{\hat{c}} \lambda^{nd-k}(T_0)\lambda^k(A \cap Z_0)$ on $(M, \mathcal{B}(M))$ is nontrivial and does not depend on $x \in U_0$, so U_0 is a small set with respect to the kernel P^n. As U_0 is a neighborhood of x^*, the point x^* is a Doeblin point with respect to P^n. $\square$

The next theorem, Theorem 6.9, summarizes the consequences of Proposition 6.6 in case x^* is accessible. It is first useful to rephrase the accessibility condition for the class of Markov chains considered here (i.e., induced by a random dynamical system).

Proposition 6.8 *A point $y \in M$ is accessible from $x \in M$ if and only if for every neighborhood U of y there exists a finite sequence $\theta_1, \ldots, \theta_n$ with $\theta_i \in \mathsf{supp}(m)$ for all i such that $F_{\theta_n} \circ \ldots \circ F_{\theta_1}(x) \in U$.*

Proof This easily follows from the definitions and the continuity of $\theta \mapsto F_\theta(x)$.
□

Recall that a point $y \in M$ is called accessible provided that it is accessible from every $x \in M$. As usual, we let Γ denote the accessible set, i.e., the set of accessible points.

Theorem 6.9 *Assume that there exists an accessible point $x^* \in M$ for which the assumptions of Proposition 6.6 hold. Then Γ has nonempty interior, P has at most one invariant probability measure μ, and $\mathsf{supp}(\mu) = \Gamma$ provided that μ exists.*

Assume in addition that for every $\theta \in \Theta$, F_θ is a diffeomorphism from M onto $F_\theta(M)$. Then $\Gamma = \overline{\mathsf{Int}(\Gamma)}$ and μ, when it exists, is absolutely continuous with respect to the Lebesgue measure on $\mathbb{R}^k$.

Proof By Proposition 6.6 and Theorem 6.2, P is ξ-irreducible. Then $\mathsf{supp}(\xi) \subset \Gamma$. The proof of Proposition 6.6 shows that ξ is (up to a multiplicative factor) the Lebesgue measure on some open subset of $\mathbb{R}^k$. Therefore its support has nonempty interior. Uniqueness of μ, when it exists, follows from Theorem 6.2 and the equality $\mathsf{supp}(\mu) = \Gamma$ follows from Proposition 5.8 (iii) (bearing in mind that P is Feller and ξ-irreducible, hence indecomposable).

If for all $\theta \in \Theta$, F_θ is a diffeomorphism, the set

$$\bigcup_{n \geq 1} \bigcup_{(\theta_1, \ldots, \theta_n) \in \mathsf{supp}(m)^n} F_{\theta_n} \circ \ldots \circ F_{\theta_1}(\mathsf{Int}(\Gamma))$$

is open and dense (by Proposition 6.8) in Γ.

The proof of absolute continuity goes as follows. Let μ be the invariant probability measure and write its Lebesgue decomposition $\mu = \mu_{ac} + \mu_s$, where μ_{ac} is absolutely continuous with respect to λ^k (written $\mu_{ac} \ll \lambda^k$) and μ_s is singular. Since $\xi \ll \lambda^k$ and $\xi \ll \mu$, the absolutely continuous part μ_{ac} is nonzero. For all $A \in \mathcal{B}(M)$,

$$\mu_{ac} P(A) = \int_\Theta \mu_{ac}(F_\theta^{-1}(A)) \, m(d\theta).$$

This shows that $\mu_{ac} P \ll \lambda^k$, because whenever $\lambda^k(A) = 0$ then $\lambda^k(F_\theta^{-1}(A)) = 0$. Thus, by uniqueness of the Lebesgue decomposition, the equality $\mu_{ac} P + \mu_s P = \mu_{ac} + \mu_s$ implies that $\mu_{ac} P(\cdot) \leq \mu_{ac}(\cdot)$. Thus $\frac{\mu_{ac}}{\mu_{ac}(M)}$ is an excessive probability measure, hence invariant. By uniqueness of the invariant probability measure, $\mu = \frac{\mu_{ac}}{\mu_{ac}(M)}$, so $\mu \ll \lambda^k$.
□

Example 6.10 (Additive Noise) Recall the setting of Exercise 3.4. We have $M = \Theta = \mathbb{R}^k$, $F : M \to M$, $F_\theta(x) := F(x) + \theta$ for $(\theta, x) \in \Theta \times M$, and $m(d\theta) = h(\theta)\, d\theta$, where $h \in L^1(d\theta)$. Assume in addition that F is C^1, which implies that $(\theta, x) \mapsto F_\theta(x)$ is C^1 as well. Finally, suppose that there is a nonempty open set $V \subset \Theta$ such that

$$\inf_{\theta \in V} h(\theta) > 0.$$

For every $x^* \in M$ and $\theta^* \in \Theta$,

$$D\varphi_{1,x^*}(\theta)|_{\theta=\theta^*} = \mathbf{1}_{k \times k},$$

where $\mathbf{1}_{k \times k}$ is the identity matrix of dimensions $(k \times k)$. Since $\mathbf{1}_{k \times k}$ has rank k, every pair $(x^*, \theta^*) \in M \times V$ satisfies the conditions of Proposition 6.6. Hence, every point $x^* \in M$ is a Doeblin point with respect to the Markov kernel $P(x, G) = m(\{\theta \in \Theta : F_\theta(x) \in G\})$.

Exercise 6.11 (Degenerate Additive Noise) Let m be a probability measure on $(\mathbb{R}, \mathcal{B}(\mathbb{R}))$ that is absolutely continuous with respect to Lebesgue measure on $\mathbb{R}$, with probability density function h. Assume further that there is a nonempty open interval $I \subset \mathbb{R}$ such that

$$\inf_{\theta \in I} h(\theta) > 0.$$

Show the following statements.

(i) Let $F : \mathbb{R}^2 \to \mathbb{R}^2$, $F = (F_1, F_2)^\top$ be a C^1-mapping and let $(x^*, \theta_1^*) \in \mathbb{R}^2 \times I$ such that

$$\partial_{x_1} F_2(F(x^*) + \theta_1^* e_1) \neq 0,$$

where $e_1 = (1, 0)^\top$. Set

$$F_\theta(x) := F(x) + \theta e_1, \quad (x, \theta) \in \mathbb{R}^2 \times \mathbb{R}.$$

Then x^* is a weak Doeblin point for the Markov kernel associated with (F, m).

(ii) Let $k \geq 2$ and let $F : \mathbb{R}^k \to \mathbb{R}^k$ be defined by

$$F(x_1, \ldots, x_k) := (x_k, x_1, x_2, \ldots, x_{k-1})^\top.$$

Set

$$F_\theta(x) := F(x) + \theta e_1, \quad (x, \theta) \in \mathbb{R}^k \times \mathbb{R},$$

where $e_1 = (1, 0, \ldots, 0)^\top \in \mathbb{R}^k$. Then any point $x^* \in \mathbb{R}^k$ is a weak Doeblin point for the Markov kernel associated with (F, m).

6.3   Random Switching Between Vector Fields

Let $E := \{1, \ldots, N\}$ be a finite set called a set of *environments*, and for each $i \in E$, let G_i be a C^∞-vector field defined on $\mathbb{R}^k$. The choice of $\mathbb{R}^k$ is made here for simplicity, but we could also assume that the G_i's are defined on a smooth k-dimensional Riemannian manifold.

By the Cauchy–Lipschitz theorem, for every $x \in \mathbb{R}^k$ the initial-value problem

$$\dot{x}(t) = G_i(x(t)),$$
$$x(0) = x$$

has a unique (local) solution $t \mapsto \Phi_i(t, x)$. We assume here that every G_i is *complete*, meaning that $t \mapsto \Phi_i(t, x)$ is defined for all $t \in \mathbb{R}$. A classical sufficient condition for completeness is that

$$\|G_i(x)\| \le a\|x\| + b, \quad \forall x \in \mathbb{R}^k,$$

for some $a, b \ge 0$. The function $\Phi_i : \mathbb{R} \times \mathbb{R}^k \to \mathbb{R}^k$ is called a *flow function*.

Let now $M \subset \mathbb{R}^k$ be a nonempty open set *positively invariant* under each Φ_i, meaning that $\Phi_i(t, M) \subset M$ for all $t \ge 0$. Consider the non-autonomous differential equation

$$\dot{Y}_t = G_{I_t}(Y_t), \tag{6.1}$$

where $t \mapsto I_t \in E$ is a *right-continuous control*, i.e.,

$$I_t = i_k \text{ for } \tau_{k-1} \le t < \tau_k, \; k \ge 1,$$

$$0 = \tau_0 < \tau_1 < \ldots < \tau_k < \tau_{k+1},$$

for some sequence $(\tau_k)_{k \ge 0}$ with $\lim_{k \to \infty} \tau_k = \infty$. Throughout this section, we shall assume that the sequence

$$\theta_1 = (\tau_1, i_1), \theta_2 = (\tau_2 - \tau_1, i_2), \ldots, \theta_k = (\tau_k - \tau_{k-1}, i_k), \ldots$$

forms a sequence of independent identically distributed random variables on $\Theta := \mathbb{R}_+ \times E$ having distribution m, where

$$m([0, t] \times \{i\}) = p_i \int_0^t \rho_i(s)\,ds, \tag{6.2}$$

$p_i > 0$ for every i, and the densities ρ_i are such that

$$\inf_{0 < s < R} \rho_i(s) > 0$$

for some $R > 0$. We also always assume that the initial value Y_0 is a random variable independent of the sequence $(\theta_k)_{k \geq 1}$.

In words, the process $Y = (Y_t)_{t \geq 0}$, with initial value Y_0 and solving the differential equation in (6.1), can be described as follows: Pick an initial pair (τ_1, i_1) at random according to m and follow the trajectory starting at Y_0 and induced by G_{i_1} for the time τ_1. Then pick a new pair (Δ_2, i_2) according to m, independent of (τ_1, i_1), and follow the trajectory starting at Y_{τ_1} and induced by G_{i_2} for the time $\Delta_2 = \tau_2 - \tau_1$. Repeating this process defines $(Y_t)_{t \geq 0}$.

The key point here is that, letting $X_n = Y_{\tau_n}$, (X_n) is a Markov chain induced by the random dynamical system (F, m), where for every $\theta = (t, i) \in \Theta$,

$$F_\theta : M \to M, \ x \mapsto \Phi_i(t, x).$$

Its kernel P is then given as

$$Pf(x) = \sum_{i \in E} p_i \int_0^\infty f(\Phi_i(t, x)) \rho_i(t) dt. \tag{6.3}$$

The following exercises give concrete examples of such a Markov chain.

Exercise 6.12 Suppose $E = \{1, 2, 3\}$, $M = \mathbb{R}$, $G_1(x) = \alpha_1$, $G_2(x) = -\alpha_2$, $G_3(x) = -\alpha_3 x$, and $\rho_i(t) = \lambda_i e^{-\lambda_i t} \mathbf{1}_{t > 0}$, $i \in E$, where α_i, λ_i are positive numbers.

Prove that the Markov kernel P associated with (F, m) admits a unique invariant probability measure. *Hint*: Use Theorem 4.31 on random contractions.

Exercise 6.13 Suppose $E = \{1, 2\}$, $M = \mathbb{R}$, $G_1(x) = \alpha_1$, $G_2(x) = -\alpha_2$, and $\rho_i(t) = \lambda_i e^{-\lambda_i t} \mathbf{1}_{t > 0}$, $i \in E$, where α_i, λ_i are positive numbers. Consider the function

$$f(t, i) := (-1)^{i+1} \alpha_i t, \quad (t, i) \in \Theta,$$

and the Borel measure

$$a(A) := m(\{\theta \in \Theta : f(\theta) \in A\}), \quad A \in \mathcal{B}(\mathbb{R}).$$

Show that the Markov kernel P associated with (F, m) satisfies

$$P(x, G) = a(\{\xi \in \mathbb{R} : x + \xi \in G\}), \quad x \in \mathbb{R}, G \in \mathcal{B}(\mathbb{R}).$$

Prove that if $p_1 \alpha_1 / \lambda_1 \neq p_2 \alpha_2 / \lambda_2$, P does not admit any invariant probability measures. *Hint*: See Example 4.19.

6.3.1 The Weak Bracket Condition

The main result in this section (Theorem 6.16) is a sufficient condition for the existence of a weak Doeblin point with respect to the Markov kernel P induced by (F, m). This condition will be formulated in terms of the Lie algebra generated by $(G_i)_{i \in E}$. The *Lie bracket* of two C^1-vector fields G and H on a nonempty open subset M of $\mathbb{R}^k$ is itself a vector field on M, defined as

$$[G, H](x) := DH(x)G(x) - DG(x)H(x), \quad x \in M.$$

Here, $DG(x)$ and $DH(x)$ denote the Jacobian matrices of G and H, respectively, evaluated at the point x. The products $DH(x)G(x)$ and $DG(x)H(x)$ are to be understood as matrix-vector products.

If Φ_G and Φ_H denote the respective flow functions of G and H, one has the alternative characterization

$$[G, H](x) = \frac{d}{dt} L(t, x)|_{t=0}, \quad x \in M, \tag{6.4}$$

where

$$L(t, x) := \Phi_H \left(-\sqrt{t}, \Phi_G \left(-\sqrt{t}, \Phi_H \left(\sqrt{t}, \Phi_G \left(\sqrt{t}, x \right) \right) \right) \right)$$

for $t \geq 0$ and $x \in M$ (see, for example, Proposition 3.b in Chapter 2 of [41]). Notice that for every fixed $x \in M$, $L(\cdot, x)$ is defined in a neighborhood of 0 because G and H are C^1.

Exercise 6.14 (Properties of Lie Brackets)

(i) Show that the Lie bracket $[\cdot, \cdot]$ is bilinear and antisymmetric, i.e., for any C^1-vector fields A, B, C and for any $\lambda \in \mathbb{R}$, one has

$$[\lambda A, C] = \lambda[A, C], \quad [A + B, C] = [A, C] + [B, C], \quad [A, B] = -[B, A].$$

Why is this enough to deduce linearity for the second argument?

(ii) To a vector field A on M, one can associate the operator on $C^\infty(M, \mathbb{R})$ that maps $f \in C^\infty(M, \mathbb{R})$ to $x \mapsto \langle A(x), \nabla f(x) \rangle$. Here, $\langle \cdot, \cdot \rangle$ denotes the Euclidean inner product on $\mathbb{R}^k$ and ∇f denotes the gradient of f. This operator is usually identified with A, so one writes Af for the image of f under the operator. Let A and B be C^2-vector fields on M. Show that

$$[A, B] = AB - BA,$$

where AB and BA should be interpreted as compositions of the operators A and B.

(iii) Use the result from (ii) to prove the *Jacobi identity*: For C^3-vector fields A, B, C, one has

$$[A, [B, C]] + [B, [C, A]] + [C, [A, B]] = 0.$$

We inductively define a sequence of families of vector fields by $G_0 := \{G_i\}_{i \in E}$ and $G_{n+1} := G_n \cup \{[G_i, V] : i \in E, V \in G_n\}$ for $n \in \mathbb{N}$. Recall that the *linear span* of a set S contained in some vector space is the set of all (finite) linear combinations of elements in S. We say that the *weak bracket condition* holds at a point $x \in M$ if the linear span of $\{V(x) : V \in \bigcup_{n \in \mathbb{N}} G_n\}$ is equal to the full space $\mathbb{R}^k$. As alluded to earlier, this condition admits an alternative formulation in terms of the Lie algebra generated by $(G_i)_{i \in E}$. The latter is defined as the smallest linear subspace $\mathcal{L}$ of the vector space of C^∞-vector fields on M that is closed under Lie brackets ($[G, H] \in \mathcal{L}$ for all $G, H \in \mathcal{L}$) and contains $(G_i)_{i \in E}$.

Exercise 6.15 Let $\mathcal{L}$ denote the Lie algebra generated by $(G_i)_{i \in E}$.

(i) Show that $G_n \subset \mathcal{L}$ for all $n \in \mathbb{N}$.
(ii) Deduce from (i) that the weak bracket condition at a point x implies that $\{V(x) : V \in \mathcal{L}\} = \mathbb{R}^k$.
(iii) Show that $\mathcal{G}$, the linear span of $\bigcup_{n \in \mathbb{N}} G_n$, is closed under Lie brackets. *Hint*: This will follow once it is shown that for every $n \in \mathbb{N}$, $A \in G_n$, and $B \in \mathcal{G}$, one has $[A, B] \in \mathcal{G}$. The Jacobi identity from Exercise 6.14 may be helpful.
(iv) Conclude that the weak bracket condition holds at a point $x \in M$ if and only if $\{V(x) : V \in \mathcal{L}\} = \mathbb{R}^k$.

We now state the main result of Sect. 6.3.1.

Theorem 6.16 *If the weak bracket condition holds at a point $x^* \in M$, then there is $n \in \mathbb{N}$ such that x^* is a Doeblin point with respect to P^n. In particular, x^* is a weak Doeblin point with respect to P.*

The proof of Theorem 6.16 relies on a slight generalization of Proposition 6.6. To state this generalization, let T be a Borel subset of $\mathbb{R}^d$ ($d \in \mathbb{N}^*$) with nonempty interior, and let E be a finite set. Let m be a probability measure on $\Theta := T \times E$, equipped with the product σ-field of $\mathcal{B}(T)$ and the power set of E. As in Sect. 6.2, the n-fold product measure $m \otimes \ldots \otimes m$ will be denoted by m^n. Let M be a nonempty open subset of $\mathbb{R}^k$, $k \in \mathbb{N}^*$, with Borel σ-field $\mathcal{B}(M)$. Let $F : \Theta \times M \to M$ be a map such that for every $i \in E$, $(t, x) \mapsto F_{(t,i)}(x)$ is C^1. For $n \in \mathbb{N}^*$, $\mathbf{i} = (i_1, \ldots, i_n) \in E^n$, and $x \in M$, let

$$\varphi^{\mathbf{i}}_{n,x} : T^n \to M, \quad (t_1, \ldots, t_n) \mapsto (F_{(t_n, i_n)} \circ \ldots \circ F_{(t_1, i_1)})(x).$$

Proposition 6.17 *Let $x^* \in M$, $n \in \mathbb{N}^*$, and $\mathbf{t}^* = (t_1^*, \ldots, t_n^*) \in \mathsf{Int}(T^n)$ such that the following conditions hold.*

(i) *There is $\mathbf{i} \in E^n$ such that the Jacobian matrix $D\varphi^{\mathbf{i}}_{n,x^*}(\mathbf{t}^*)$ has rank k;*

(ii) *There is a neighborhood $V \subset \text{Int}(T^n)$ of $\mathbf{t}^*$ such that $m^n((\cdot \cap V) \times \{\mathbf{i}\})$ is absolutely continuous with respect to $\lambda^{nd}(\cdot \cap V)$. The corresponding probability density function ρ can be chosen such that*

$$c := \inf_{\mathbf{t} \in V} \rho(\mathbf{t}) > 0.$$

Under these conditions, x^* is a Doeblin point with respect to P^n, and in particular a weak Doeblin point with respect to P.

Exercise 6.18 Prove Proposition 6.17. *Hint:* The proof of Proposition 6.6 can almost be repeated verbatim.

The setting of randomly switched vector fields introduced at the beginning of Sect. 6.3 is clearly covered by the more general setting of Proposition 6.17, with $T = \mathbb{R}_+$ and $d = 1$. The proof of Theorem 6.16 therefore reduces to checking conditions 1 and 2 of Proposition 6.17. While condition 2 follows almost immediately from the definition of m, establishing condition 1 requires a link between the weak bracket condition and the full-rank condition on the Jacobian matrix of $\varphi^i_{n,x}$. This link is provided by the following result from geometric control theory, which is implied by Theorem 1 of Chapter 3 in [41]. To help the reader understand this result, we give its proof.

Lemma 6.19 *Under the assumptions of Theorem 6.16 and for $1 \le j \le k$, the following statement holds: For every $\varepsilon > 0$ there are $\mathbf{i} \in E^j$ and $\mathbf{t}^* \in (0, \varepsilon)^j$ such that $D\varphi^{\mathbf{i}}_{j,x^*}(\mathbf{t}^*)$ has rank j.*

Proof We prove Lemma 6.19 by induction. In the base case $j = 1$, the weak bracket condition at x^* implies that there is $i \in E$ such that $G_i(x^*) \neq 0$. Then for every $\varepsilon > 0$ there is $t^* \in (0, \varepsilon)$ such that $G_i(\Phi_i(t^*, x^*)) \neq 0$. Since

$$\varphi^i_{1,x^*}(t^*) = F_{(t^*,i)}(x^*) = \Phi_i(t^*, x^*),$$

one has

$$D\varphi^i_{1,x^*}(t^*) = G_i(\Phi_i(t^*, x^*)),$$

which has rank 1.

In the induction step, assume that the statement holds for some $1 \le j < k$, and let $\varepsilon > 0$. Since the weak bracket condition holds at x^*, it also holds in an open neighborhood $M^* \subset M$ of x^*. There is no loss of generality in assuming that ε is so small that $\varphi^{\mathbf{i}}_{j,x^*}(\mathbf{t}) \in M^*$ for every $\mathbf{i} \in E^j$ and $\mathbf{t} \in (0, \varepsilon)^j$. By induction hypothesis, there are $\mathbf{i} \in E^j$ and $\mathbf{t}^* \in (0, \varepsilon)^j$ such that $D\varphi^{\mathbf{i}}_{j,x^*}(\mathbf{t}^*)$ has rank j. Since a full rank is preserved under small perturbations of the matrix entries, there is an open neighborhood N of $\mathbf{t}^*$ in $(0, \varepsilon)^j$ such that $D\varphi^{\mathbf{i}}_{j,x^*}$ has rank j on N. The mapping $\varphi^{\mathbf{i}}_{j,x^*}$ is then a differentiable map between the manifolds N and M, and

$D\varphi^{\mathbf{i}}_{j,x*}$ has constant rank j on N. By the constant-rank theorem (see, e.g., Theorem 2.b of Chapter 2 in [41]), there is an open neighborhood U of $\mathbf{t}^*$ in N such that $S := \varphi^{\mathbf{i}}_{j,x*}(U)$ is an embedded submanifold of M of dimension j.

We call a vector field V *tangent* to S if for every $y \in S$, $V(y)$ is a vector in $T_y S$, the tangent space with respect to S at the point y. We will now show that there is at least one vector field G_i, $i \in E$, that is not tangent to S.

Assume this is not the case, i.e., G_i is tangent to S for every $i \in E$. The set of vector fields tangent to S is clearly closed under linear combinations. It is also closed under the Lie bracket operation because of the flow-based characterization of the Lie bracket in (6.4) and the fact that the flow of a vector field tangent to S stays in S for t in a nonempty open interval around 0 (see Proposition 1 of Chapter 2 in [41]). This shows that every vector field in $\mathcal{L}$, the Lie algebra generated by $(G_i)_{i \in E}$, is tangent to S. Fix an arbitrary point $y \in S$. The submanifold S was defined in such a way that the weak bracket condition holds at every point in S and in particular at y. Since $V(y) \in T_y S$ for every $V \in \mathcal{L}$, the tangent space $T_y S$ has dimension k, which is strictly larger than j. This contradicts the fact that S has dimension j.

Let $y \in S$ and $i_{j+1} \in E$ such that $G_{i_{j+1}}(y) \notin T_y S$. There is $\hat{\mathbf{t}} \in U$ such that $y = \varphi^{\mathbf{i}}_{j,x*}(\hat{\mathbf{t}})$. Then

$$
\begin{aligned}
D\varphi^{\mathbf{i},i_{j+1}}_{j+1,x*}(\hat{\mathbf{t}}, 0) &= D_{(t_1,\dots,t_{j+1})}\Phi_{i_{j+1}}(t_{j+1}, \varphi^{\mathbf{i}}_{j,x*}(t_1,\dots,t_j))|_{(t_1,\dots,t_j)=\hat{\mathbf{t}}, t_{j+1}=0} \\
&= \left(D\varphi^{\mathbf{i}}_{j,x*}(\hat{\mathbf{t}}), G_{i_{j+1}}(\varphi^{\mathbf{i}}_{j,x*}(\hat{\mathbf{t}})) \right) = \left(D\varphi^{\mathbf{i}}_{j,x*}(\hat{\mathbf{t}}), G_{i_{j+1}}(y) \right).
\end{aligned}
$$

Since $\hat{\mathbf{t}} \in N$, the matrix $D\varphi^{\mathbf{i}}_{j,x*}(\hat{\mathbf{t}})$ has rank j. As a result, the columns of $D\varphi^{\mathbf{i}}_{j,x*}(\hat{\mathbf{t}})$ are j linearly independent elements of $T_y S$. Since the $(j + 1)$st column of $D\varphi^{\mathbf{i},i_{j+1}}_{j+1,x*}(\hat{\mathbf{t}}, 0)$ is not contained in $T_y S$, it follows that $D\varphi^{\mathbf{i},i_{j+1}}_{j+1,x*}(\hat{\mathbf{t}}, 0)$ has rank $(j + 1)$. Again by virtue of the fact that having full rank is preserved under small perturbations of the matrix entries, it follows that for $t \in (0, \varepsilon)$ sufficiently small, $D\varphi^{\mathbf{i},i_{j+1}}_{j+1,x*}(\hat{\mathbf{t}}, t)$ has rank $(j + 1)$. □

We are now ready to prove Theorem 6.16.

Proof *(of Theorem 6.16)* Let $x^* \in M$ be a point where the weak bracket condition holds. By Lemma 6.19, there are $\mathbf{i} \in E^k$ and $\mathbf{t}^* \in V := (0, R)^k$ such that $D\varphi^{\mathbf{i}}_{k,x*}(\mathbf{t}^*)$ has rank k.

For Borel sets $A_1, \dots, A_k \subset (0, R)$ and $A := A_1 \times \dots \times A_k$, we have

$$
m^k\left(\prod_{l=1}^{k} A_l \times \{i_l\} \right) = \prod_{l=1}^{k} m(A_l \times \{i_l\}) = \int_A \rho(\mathbf{t})\, d\mathbf{t},
$$

where

$$\rho(\mathbf{t}) := \prod_{l=1}^{k} p_{i_l} \rho_{i_l}(t_l)$$

and thus $\inf_{t \in V} \rho(\mathbf{t}) > 0$. The theorem then follows from Proposition 6.17.   □

The following proposition is implied by Proposition 6.8 and the definition of a right-continuous control at the beginning of Sect. 6.3.

Proposition 6.20 *A point $y \in M$ is accessible from $x \in M$ for P (given by 6.3) if and only if for every neighborhood U of y there exists a right-continuous control $j : \mathbb{R}_+ \to E$ such that the solution $t \mapsto x(t)$ to the initial-value problem*

$$\dot{x}(t) = G_{j(t)}(x(t))$$

$$x(0) = x$$

meets U. That is, $x(t) \in U$ for some $t \geq 0$.

In the proof of Theorem 6.16 it was shown that if the weak bracket condition holds at a point $x^* \in M$, then the assumptions of Proposition 6.17 are satisfied for $n = k, T = \mathbb{R}_+$, and $V = (0, R)^k$. Furthermore, by our assumptions on $(G_i)_{i \in E}$, F_θ is a C^1-diffeomorphism (even a C^∞-diffeomorphism) for every $\theta \in \Theta$. In analogy to Theorem 6.9, one obtains the following corollary. As usual, we let Γ denote the set of points that are accessible from every point in M.

Corollary 6.21 *If the weak bracket condition holds at an accessible point $x^* \in M$, then $\Gamma = \overline{\mathsf{Int}(\Gamma)}$ and P has at most one invariant probability measure μ. When it exists, μ is absolutely continuous with respect to λ^k and $\mathsf{supp}(\mu) = \Gamma$.*

6.4 Piecewise Deterministic Markov Processes

In this section, we keep the notation of the preceding Sect. 6.3 but restrict our attention to the specific case where the densities $\rho_i, i \in E$, appearing in the definition of the measure m (see (6.2)) are exponential, i.e.,

$$\rho_i(t) = \lambda_i e^{-\lambda_i t} \mathbf{1}_{t > 0}$$

with $\lambda_i > 0$.

We shall consider certain properties of the *joint continuous-time* process $Z_t = (Y_t, I_t)$. Such a process is sometimes called in the literature a *piecewise deterministic Markov process*, in short a PDMP.

For all $f : M \times E \to \mathbb{R}$ bounded and measurable and for all $t \geq 0$, we let

$$P_t f(x, i) = \mathsf{E}(f(Z_t)|Z_0 = (x, i)). \qquad (6.5)$$

Remark 6.22 Alternatively, one can define P_t as follows. Given $(x, i) \in M \times E$, let $(Z_t^{x,i}) = (Y_t^{x,i}, I_t^i)$ denote the continuous-time process characterized by

$$\dot{Y}_t^{x,i} = G_{I_t^i}(Y_t^{x,i}), \; Y_0^{x,i} = x,$$

where (I_t^i) is defined like (I_t) with the exception that θ_1 has law $\rho_i(t)dt \otimes \delta_i$ instead of m. Then

$$P_t f(x, i) = \mathsf{E}(f(Z_t^{x,i})).$$

Proposition 6.23 *The semigroup* $\{P_t\}_{t \geq 0}$ *is weakly Feller and* $(Z_t)_{t \geq 0}$ *is a Markov process with semigroup* $\{P_t\}_{t \geq 0}$.

Exercise 6.24 Prove Proposition 6.23. (*Hint:* Use the memoryless property of the exponential distribution: $\mathsf{P}(\tau_1 > t + s|\tau_1 > t) = \mathsf{P}(\tau_1 > s)$.) Explain why (*ii*) fails to hold if the ρ_i's are not exponential.

Exercise 6.25 Let $C_c^1(M \times E)$ denote the set of maps $f : M \times E \to \mathbb{R}, (x, i) \mapsto f(x, i)$ that are C^1 in x and have compact support. For $f \in C_c^1(M \times E)$, we let $\nabla f(x, i)$ denote the gradient of $x \mapsto f(x, i)$. Let $\mathcal{D}, \mathcal{L} : C_c^1(M \times E) \to B(M \times E)$ be the (unbounded) operators defined by

$$\mathcal{D}f(x, i) = \langle \nabla f(x, i), G_i(x) \rangle,$$

and

$$\mathcal{L}f(x, i) = \mathcal{D}f(x, i) + \lambda_i \sum_{j \in E} p_j(f(x, j) - f(x, i)).$$

Prove that for all $f \in C_c^1(M \times E)$,

$$\lim_{t \to 0} \frac{P_t f(x, i) - f(x, i)}{t} = \mathcal{L}f(x, i)$$

and that the convergence is uniform in x. In the language of continuous-time Markov processes, $\mathcal{L}$ is called the *infinitesimal generator* of the Markov semigroup $\{P_t\}$.

6.4.1 Invariant Measures

The following result relates invariant probability measures of the discrete kernel P (given by (6.3)) to invariant probability measures of $\{P_t\}$.

Theorem 6.26 *Let* $\mathsf{Inv}(P)$ *(respectively* $\mathsf{Inv}(\{P_t\})$*) be the set of invariant probability measures for* P *(respectively* $\{P_t\}$*). Let* $c = \frac{1}{\sum_{j \in E} p_j/\lambda_j}$.

(i) *If* $\mu \in \mathsf{Inv}(P)$, *then* $\hat{\mu} \in \mathsf{Inv}(\{P_t\})$, *where* $\hat{\mu}$ *is defined by*

$$\hat{\mu}(A \times \{i\}) = c\, p_i \int_0^\infty \mu(\Phi_i(-t, A)) e^{-\lambda_i t}\, dt;$$

in this formula we think of μ *as a measure on* $\mathbb{R}^k$ *that only charges* M;
(ii) *If* $v \in \mathsf{Inv}(\{P_t\})$, *then* $\check{v} \in \mathsf{Inv}(P)$, *where* $\check{v}$ *is defined by*

$$\check{v}(A) = \frac{1}{c} \sum_{i \in E} \lambda_i v(A \times \{i\});$$

(iii) *The mappings* $\mathsf{Inv}(P) \to \mathsf{Inv}(\{P_t\}) : \mu \mapsto \hat{\mu}$ *and* $\mathsf{Inv}(\{P_t\}) \to \mathsf{Inv}(P) : v \mapsto \check{v}$, *are inverse to each other;*
(iv) $\mathsf{supp}(\hat{\mu}) = \mathsf{supp}(\mu) \times E$.

Proof

(i) Let μ be a Borel probability measure on M. Then, for all $f \in B(M \times E)$,

$$\hat{\mu}(f) = \frac{\mathbb{E}_{\mu \otimes p}(\int_0^{\tau_1} f(Z_s)\, ds)}{\mathbb{E}_{\mu \otimes p}(\tau_1)}, \tag{6.6}$$

where $\mu \otimes p$ stands for the product measure $\mu \otimes p = \sum_i p_i \mu(dx) \delta_i$. Indeed,

$$\mathbb{E}_{\mu \otimes p}\left(\int_0^{\tau_1} f(Z_s)\, ds\right) = \sum_{i \in E} p_i \int_M \mathbb{E}\left(\int_0^{\tau_1} f(Z_s^{x,i})\, ds\right) \mu(dx)$$

$$= \sum_{i \in E} p_i \int_M \int_0^\infty \int_0^t f(\Phi_i(s, x), i)\, ds\, \lambda_i e^{-\lambda_i t}\, dt\, \mu(dx)$$

$$= \sum_{i \in E} p_i \int_M \int_0^\infty f(\Phi_i(t, x), i) e^{-\lambda_i t}\, dt\, \mu(dx) = c^{-1} \hat{\mu}(f),$$

where we used integration by parts. This equality applied with $f \equiv 1$ gives $\mathbb{E}_{\mu \otimes p}(\tau_1) = c^{-1}$, so the formula in (6.6) follows. If now $\mu \in \mathsf{Inv}(P)$ and Z_0 has distribution $\mu \otimes p$, then Z_{τ_1} has the same distribution. A continuous-time version of Exercise 4.24 proves that $\hat\mu$ lies in $\mathsf{Inv}(\{P_t\})$. More precisely, for every $t > 0$ and $f \in B(M \times E)$, Proposition 6.23 (ii) yields

$$\mathbb{E}_{\mu\otimes p}\left(\int_0^{\tau_1} P_t f(Z_s)\,ds\right) = \mathbb{E}_{\mu\otimes p}\left(\int_0^\infty \mathbb{E}_{\mu\otimes p}(f(Z_{s+t})\mathbf{1}_{s<\tau_1}|\mathcal{F}_s)\,ds\right)$$

$$= \mathbb{E}_{\mu\otimes p}\left(\int_0^\infty f(Z_{s+t})\mathbf{1}_{s<\tau_1}\,ds\right)$$

$$= \mathbb{E}_{\mu\otimes p}\left(\int_t^{t+\tau_1} f(Z_s)\,ds\right) = \mathbb{E}_{\mu\otimes p}\left(\int_0^{\tau_1} f(Z_s)\,ds\right).$$

Here the last equality comes from the fact that

$$\mathbb{E}_{\mu\otimes p}\left(\int_{\tau_1}^{t+\tau_1} f(Z_s)\,ds - \int_0^t f(Z_s)\,ds\right) = 0$$

because $(Z_t)_{t\geq0}$ and $(Z_{\tau_1+t})_{t\geq0}$ have the same distribution. In light of (6.6), we have thus shown that $\hat\mu(P_t f) = \hat\mu(f)$ for every $f \in B(M \times E)$ and $t > 0$, hence $\hat\mu \in \mathsf{Inv}(\{P_t\})$.

(ii) Let now $\nu \in \mathsf{Inv}(\{P_t\})$. We shall show that $\check\nu \in \mathsf{Inv}(P)$.

Let $K_t, K, \lambda, \lambda^{-1} : B(M \times E) \to B(M \times E)$ and $Q : B(M \times E) \to B(M)$ be the (bounded) operators respectively defined by

$$K_t f(x, i) = f(\Phi_i(t, x), i),$$

$$Kf(x, i) = \int_0^\infty \lambda_i e^{-\lambda_i t} K_t f(x, i)\,dt,$$

$$\lambda f(x, i) = \lambda_i f(x, i), \quad \lambda^{-1} f(x, i) = \lambda_i^{-1} f(x, i),$$

and

$$Qf(x) = \sum_{j\in E} p_j f(x, j).$$

Let $\mathcal{D}$ and $\mathcal{L}$ be the unbounded operators on $C_c^1(M \times E)$ as defined in Exercise 6.25.

Let $f \in C_c^1(M \times E)$. Then

$$\mathcal{L}f = (\mathcal{D} - \lambda)f + \lambda Qf. \tag{6.7}$$

One has

$$\frac{dK_t f}{dt} = \mathcal{D}K_t f = K_t \mathcal{D} f, \qquad (6.8)$$

where, for the second inequality, we used

$$G_i(\Phi_i(t, x)) = D_x \Phi_i(t, x)|_{x=\Phi_i(t,x)} G_i(x).$$

Furthermore,

$$\int_0^\infty \frac{dK_t f(x, i)}{dt} e^{-\lambda_i t} \, dt = -f(x, i) + Kf(x, i) \qquad (6.9)$$

with integration by parts. The relations in (6.8) and (6.9) together with

$$\nabla K f(x, i) = \int_0^\infty \lambda_i e^{-\lambda_i t} \nabla K_t f(x, i) \, dt$$

justified by $f \in C_c^1(M \times E)$ lead to the identity

$$K(\mathcal{D} - \lambda)f = (\mathcal{D} - \lambda)Kf = -\lambda f.$$

Thus, with (6.7),

$$\mathcal{L}Kf = (\mathcal{D} - \lambda)Kf + \lambda QKf = -\lambda f + \lambda QKf. \qquad (6.10)$$

Let now $f \in C_c^1(M)$. We can see f as an element of $C_c^1(M \times E)$ by setting $f(x, i) = f(x)$. For such f the identity in (6.10) reads

$$\mathcal{L}Kf(x, i) = \lambda_i(-f(x) + Pf(x)).$$

Since $v \in \mathsf{Inv}(\{P_t\})$, $vP_t(Kf) = v(Kf)$ for all t, and consequently, by dominated convergence, $v\mathcal{L}Kf = 0$ (see Exercise 6.25). Hence $\check{v}f = \check{v}Pf$ with

$$\check{v}(dx) = \frac{1}{\sum_j \lambda_j v(M \times \{j\})} \sum_i \lambda_i v(dx \times \{i\}).$$

This proves that $\check{v} \in \mathsf{Inv}(P)$ (use Remark 4.17 with $\mathcal{C} = C_c^1(M)$). Finally, observe that the equation $v\mathcal{L}f = 0$ applied to $f(x, i) = f(i)$ leads to

$$v(M \times \{i\}) = \frac{p_i}{\lambda_i} \sum_{k \in E} \lambda_k v(M \times \{k\}) = \frac{p_i}{\lambda_i} c,$$

where we used that $\sum_{i \in E} \nu(M \times \{i\}) = 1$. Hence

$$\check{\nu}(dx) = \frac{1}{c} \sum_i \lambda_i \nu(dx \times \{i\}) = \check{\nu}(dx).$$

(iii) For $\mu \in \mathsf{Inv}(P)$ and $f \in B(M \times E)$, $\hat{\mu}(f) = c\mu(Q\frac{1}{\lambda}Kf)$. For $\nu \in \mathsf{Inv}(\{P_t\})$
        and $f \in B(M)$, $\check{\nu}(f) = \frac{1}{c}\nu(\lambda f)$. Thus, for $f \in B(M)$,

$$\check{\hat{\mu}}(f) = \frac{1}{c}\hat{\mu}(\lambda f) = \mu(QKf) = \mu(Pf) = \mu(f)$$

by P-invariance. This proves that $\check{\hat{\mu}} = \mu$. Conversely, for $\nu \in \mathsf{Inv}(\{P_t\})$ and
$f \in C_c^1(M \times E)$,

$$\nu(f) = \nu(\lambda 1/\lambda f) = \nu(\lambda QK(1/\lambda f)) = c\check{\nu}(QK1/\lambda f) = c\check{\nu}(Q\frac{1}{\lambda}Kf) = \hat{\check{\nu}}(f),$$

where the second equality follows from $\nu\mathcal{L}Kf = 0$ and identity (6.10). Thus
$\nu = \hat{\check{\nu}}$.

(iv) This last assertion immediately follows from the other three.

$\square$

Remark 6.27 By Corollary 6.21 and Theorem 6.26, whenever there exists an
accessible point for P (see Proposition 6.20) at which the weak bracket condition
holds, $\{P_t\}_{t \geq 0}$ has at most one invariant probability.

6.4.2 The Strong Bracket Condition

We now define a strengthening of the weak bracket condition. Let $G_0' := \{G_i - G_j : i, j \in E\}$ and $G_{n+1}' := G_n' \cup \{[G_i, V] : i \in E, V \in G_n'\}$ for $n \in \mathbb{N}$. We
say that the *strong bracket condition* holds at a point $x \in M$ if the linear span of
$\{V(x) : V \in \bigcup_{n \in \mathbb{N}} G_n'\}$ is equal to the full space $\mathbb{R}^k$. Clearly the strong bracket
condition implies the weak one.

Exercise 6.28 Let $G_1'' := \{[G_i, G_j] : i, j \in E\}$ and $G_{n+1}'' := G_n'' \cup \{[G_i, V] : i \in E, V \in G_n''\}$ for $n \in \mathbb{N}^*$.

(i) Show that every vector field in $(\bigcup_{n \in \mathbb{N}} G_n') \setminus G_0'$ can be written as a linear
       combination of vector fields in $\bigcup_{n \in \mathbb{N}^*} G_n''$.
(ii) Let V be a vector field in the linear span of $\bigcup_{n \in \mathbb{N}} G_n'$. Show that there exist real
        numbers $(\alpha_i)_{i \in E}$ with $\sum_{i \in E} \alpha_i = 0$ and a vector field W in the linear span of

$\bigcup_{n\in\mathbb{N}^*} G_n''$ such that

$$V = W + \sum_{i\in E} \alpha_i G_i.$$

Exercise 6.29 Given a vector field V on M, define the vector fields $\mathbf{0}\oplus V$ and $\mathbf{1}\oplus V$ on $\mathbb{R} \times M$ by

$$(\mathbf{0}\oplus V)(r,x) = (0, V(x)) \quad \text{and} \quad (\mathbf{1}\oplus V)(r,x) = (1, V(x)).$$

Let $U_1'' = \{[\mathbf{1}\oplus G_i, \mathbf{1}\oplus G_j] : i, j \in E\}$ and $U_{n+1}'' = U_n'' \cup \{[\mathbf{1}\oplus G_i, V] : i \in E, V \in U_n''\}$ for $n \in \mathbb{N}^*$. Assume that V is contained in the linear span of $\bigcup_{n\in\mathbb{N}^*} G_n''$, which was introduced in Exercise 6.28. Show that $\mathbf{0}\oplus V$ lies in the linear span of $\bigcup_{n\in\mathbb{N}^*} U_n''$.

Theorem 6.30 *If the strong bracket condition holds at a point* $x^* \in M$, *then, for every* $i \in E$ *and* $t > 0$, (x^*, i) *is a Doeblin point with respect to* P_t.

Proof Let $x^* \in M$ be a point where the strong bracket condition holds, let $t > 0$, and let $i \in E$. Continuity of the vector fields $(G_j)_{j\in E}$ and their Lie brackets implies that x^* admits an open neighborhood U such that the strong bracket condition holds at every point in U. Let $\varepsilon_1 \in (0, t)$ be so small that $\Phi_i(t_0, \overset{*}{x}) \in U$ for every $t_0 \in (0, \varepsilon_1)$. Fix $t_0^* \in (0, \varepsilon_1)$ and set $y^* := \Phi_i(t_0^*, x^*)$, where now the strong bracket condition holds as well. For $d \in \mathbb{N}^*$ and $s > 0$, set

$$\Delta_{d,s} = \{(t_1, \ldots, t_d) \in (0, \infty)^d : t_1 + \ldots + t_d < s\}.$$

For $\mathbf{i} = (i_1, \ldots, i_{d+1}) \in E^{d+1}$, define the functions

$$F_{\mathbf{i}}^{d,s} : \Delta_{d,s} \times M \to M,$$

$$((t_1, \ldots, t_d), x) \mapsto \Phi_{i_{d+1}}(s - (t_1 + \ldots + t_d), \varphi_{d,x}^{(i_1,\ldots,i_d)}(t_1, \ldots, t_d)),$$

and

$$\psi_{\mathbf{i},x}^{d,s} : \Delta_{d,s} \to M, \quad (t_1, \ldots, t_d) \mapsto F_{\mathbf{i}}^{d,s}((t_1, \ldots, t_d), x).$$

The proof of Theorem 6.30 is organized as follows: We first show that there exists a sequence of indices $\mathbf{i} = (i_1, \ldots, i_{k+1}) \in E^{k+1}$ and $\mathbf{s}^* \in \Delta_{k,t-t_0^*}$ such that $D\psi_{\mathbf{i},y^*}^{k,t-t_0^*}(\mathbf{s}^*)$ has rank k. Then we show that x^* is a Doeblin point with respect to Q, the Markov kernel associated with the random dynamical system $(F_{(i,i)}^{k+1,t}, q)$, where q is the normalized Lebesgue measure $q(\cdot) = \lambda^{k+1}(\cdot)/\lambda^{k+1}(\Delta_{k+1,t})$. From this we deduce that (x^*, i) is a Doeblin point with respect to P_t.

Let us begin by showing the existence of $\mathbf{i}$ and $\mathbf{s}^*$ such that $D\psi_{\mathbf{i},y^*}^{k,t-t_0^*}(\mathbf{s}^*)$ has rank k. On $\mathbb{R} \times M$, consider the vector fields $(\mathbf{1}\oplus G_j)_{j\in E}$ defined as in Exercise 6.29.

We claim that the strong bracket condition at y^* with respect to $(G_j)_{j\in E}$ implies the weak bracket condition at (r, y^*) with respect to $(\mathbf{1} \oplus G_j)_{j\in E}$ for every $r \in \mathbb{R}$.

To see this, let $r \in \mathbb{R}$, $\mathbf{v} = (v_1, \mathbf{v}_k) \in \mathbb{R} \times \mathbb{R}^k$, and $i^* \in E$. Since the strong bracket condition holds at y^*, there exists a vector field V in the linear span of $\bigcup_{n\in\mathbb{N}} G'_n$ such that

$$\mathbf{v}_k - v_1 G_{i^*}(y^*) = V(y^*).$$

By Exercise 6.28 (ii), there exist real numbers $(\alpha_j)_{j\in E}$ with $\sum_{j\in E} \alpha_j = 0$ and a vector field W in the linear span of $\bigcup_{n\geq 1} G''_n$ (defined in Exercise 6.28) such that

$$V = W + \sum_{j\in E} \alpha_j G_j.$$

Then, by Exercise 6.29, $\mathbf{0} \oplus W$ lies in the linear span of $\bigcup_{n\in\mathbb{N}^*} U''_n$ (defined in Exercise 6.29). Let $U_0 = \{\mathbf{1} \oplus G_j : j \in E\}$ and $U_{n+1} = U_n \cup \{[\mathbf{1} \oplus G_j, V] : j \in E, V \in U_n\}$ for $n \in \mathbb{N}$. It is easy to check that $U''_n \subset U_n$ for every $n \in \mathbb{N}^*$, so $\mathbf{0} \oplus W$ lies in the linear span of $\bigcup_{n\in\mathbb{N}} U_n$. Now we can write

$$\mathbf{v} = (v_1, \mathbf{v}_k) = (0, \mathbf{v}_k - v_1 G_{i^*}(y^*)) + v_1(\mathbf{1} \oplus G_{i^*})(r, y^*)$$

$$= (\mathbf{0} \oplus W)(r, y^*) + \sum_{j\in E} \alpha_j(\mathbf{1} \oplus G_j)(r, y^*) + v_1(\mathbf{1} \oplus G_{i^*})(r, y^*),$$

where we used that $\sum_{j\in E} \alpha_j = 0$. This proves that $\mathbf{v}$ lies in the linear span of $\bigcup_{n\in\mathbb{N}} U_n$. Accordingly, the weak bracket condition with respect to $(\mathbf{1} \oplus G_j)_{j\in E}$ holds at (r, y^*).

Let $(\tilde{\Phi}_j)_{j\in E}$ denote the flow functions associated with the vector fields $(\mathbf{1} \oplus G_j)_{j\in E}$ on $\tilde{M} := \mathbb{R} \times M$. Define the maps

$$\tilde{F} : \Theta \times \tilde{M} \to \tilde{M}, \quad ((s, j), (r, x)) \mapsto \tilde{F}_{(s,j)}(r, x) := \tilde{\Phi}_j(s, (r, x))$$

and

$$\tilde{\varphi}^{\mathbf{i}}_{n,(r,x)} : \mathbb{R}^n_+ \to \tilde{M}, \quad (t_1, \dots, t_n) \mapsto (\tilde{F}_{(t_n,i_n)} \circ \dots \circ \tilde{F}_{(t_1,i_1)})(r, x)$$

for $n \in \mathbb{N}^*$, $\mathbf{i} = (i_1, \dots, i_n) \in E^n$, and $(r, x) \in \tilde{M}$. Fix $\varepsilon \in (0, \varepsilon_1 \wedge \frac{t-t_0^*}{k+1})$. Since the weak bracket condition with respect to $(\mathbf{1} \oplus G_j)_{j\in E}$ holds at (r, y^*), Theorem 6.19 implies that there are $\mathbf{i} \in E^{k+1}$ and $\mathbf{s}^{**} = (t_1^*, \dots, t_{k+1}^*) \in (0, \varepsilon)^{k+1}$ such that $D\tilde{\varphi}^{\mathbf{i}}_{k+1,(r,y^*)}(\mathbf{s}^{**})$ has rank $(k+1)$.

Set $\mathbf{s}^* = (t_1^*, \ldots, t_k^*)$ and $\tau = t_0^* + t_1^* + \ldots + t_{k+1}^*$. Then $\mathbf{s}^* \in \Delta_{k,t-t_0^*}$ and

$$D\psi_{\mathbf{i},y^*}^{k,t-t_0^*}(\mathbf{s}^*) = D\Phi_{i_{k+1}}(t - \tau, \varphi_{k+1,y^*}^{\mathbf{i}}(\mathbf{s}^{**}))D\varphi_{k+1,y^*}^{\mathbf{i}}(\mathbf{s}^{**}) \begin{pmatrix} \mathbf{1}_{k\times k} \\ -1 \cdots -1 \end{pmatrix}.$$

Since $\Phi_{i_{k+1}}(t - \tau, \cdot)$ is a diffeomorphism, the matrix $D\Phi_{i_{k+1}}(t - \tau, \varphi_{k+1,y^*}^{\mathbf{i}}(\mathbf{s}^{**}))$ is invertible. We now show that

$$D\varphi_{k+1,y^*}^{\mathbf{i}}(\mathbf{s}^{**}) \begin{pmatrix} \mathbf{1}_{k\times k} \\ -1 \cdots -1 \end{pmatrix} \tag{6.11}$$

is invertible as well and thus that $D\psi_{\mathbf{i},y^*}^{k,t-t_0^*}(\mathbf{s}^*)$ has rank k. To obtain a contradiction, suppose that the matrix in (6.11) is not invertible. Denoting the columns of $D\varphi_{k+1,y^*}^{\mathbf{i}}(\mathbf{s}^{**})$ by $a_1, \ldots, a_{k+1}$, the matrix in (6.11) becomes $(a_1 - a_{k+1}, \ldots, a_k - a_{k+1})$, so there exist $j \in \{1, \ldots, k\}$ and real numbers $(\beta_l)_{l\in\{1,\ldots,k\}\setminus\{j\}}$ such that

$$a_j - a_{k+1} = \sum_{l\in\{1,\ldots,k\}\setminus\{j\}} \beta_l(a_l - a_{k+1}).$$

Then

$$\begin{pmatrix} 1 \\ a_j \end{pmatrix} = \sum_{l\in\{1,\ldots,k\}\setminus\{j\}} \beta_l \begin{pmatrix} 1 \\ a_l \end{pmatrix} + \left(1 - \sum_{l\in\{1,\ldots,k\}\setminus\{j\}} \beta_l\right) \begin{pmatrix} 1 \\ a_{k+1} \end{pmatrix}.$$

Since

$$D\tilde{\varphi}_{k+1,(r,y^*)}^{\mathbf{i}}(\mathbf{s}^{**}) = \begin{pmatrix} 1 \cdots 1 \\ D\varphi_{k+1,y^*}^{\mathbf{i}}(\mathbf{s}^{**}) \end{pmatrix} = \begin{pmatrix} 1 \cdots 1 \\ a_1 \cdots a_{k+1} \end{pmatrix},$$

this implies that $D\tilde{\varphi}_{k+1,(r,y^*)}^{\mathbf{i}}(\mathbf{s}^{**})$ has rank strictly less than $(k+1)$, a contradiction.

Now we show that x^* is a Doeblin point with respect to Q, the Markov kernel associated with $(F_{(i,\mathbf{i})}^{k+1,t}, q)$. To do so, we will apply Proposition 6.17 with $\Delta_{k+1,t}$, $\{1\}$, q, $F_{(i,\mathbf{i})}^{k+1,t}$, and $\psi_{(i,\mathbf{i}),x}^{k+1,t}$ playing the roles of T, E, m, F, and $\varphi_{1,x}^1$, respectively. Since the finite set $\{1\}$ consists of a single element, we may identify Θ from Proposition 6.17 with $T = \Delta_{k+1,t}$. The measure q is clearly absolutely continuous with respect to λ^{k+1} and has a constant probability density function. To be able to invoke Proposition 6.17, it is then enough to show that for $\mathbf{t}^* := (t_0^*, \mathbf{s}^*)$, $D\psi_{(i,\mathbf{i}),x^*}^{k+1,t}(\mathbf{t}^*)$ has rank k. But this follows from

$$D_{(t_1,\ldots,t_k)}\psi_{(i,\mathbf{i}),x^*}^{k+1,t}(\mathbf{t}^*) = D\psi_{\mathbf{i},y^*}^{k,t-t_0^*}(\mathbf{s}^*)$$

and the fact that the matrix on the right-hand side has rank k, as established in the second step of the proof.

To complete the proof of Theorem 6.30, we argue as follows. Since x^* is a Doeblin point with respect to Q, there exist a neighborhood $B \subset M$ of x^* and a nonzero Borel measure ξ on M such that

$$Q(x, A) \geq \xi(A), \quad \forall x \in B, \ A \in \mathcal{B}(M).$$

Define the event

$$C = \{\tau_{k+1} < t < \tau_{k+2}, I_0 = i, I_{\tau_l} = i_l \text{ for } 1 \leq l \leq k+1\}.$$

For every $x \in B$, $A \in \mathcal{B}(M)$, and $j \in E$,

$$P_t((x, i), A \times \{j\}) = \mathbb{P}(Z_t^{x,i} \in A \times \{j\}|C)\mathsf{P}(C) = \delta_{i_{k+1}}(j)\mathbb{P}(Y_t^{x,i} \in A|C)\mathsf{P}(C).$$

Let $(T_0, \ldots, T_{k+1})$ be independent random variables living on some probability space with probability measure $\mathbb{P}$ such that T_0 has probability density function ρ_i and T_l has probability density function ρ_{i_l} for $1 \leq l \leq k+1$. Set

$$R = \{T_0 + \ldots + T_k < t \leq T_0 + \ldots + T_{k+1}\}.$$

Then

$$P_t((x, i), A \times \{j\}) = \delta_{i_{k+1}}(j) \, \mathbb{P}(F_{(i,i)}^{k+1,t}((T_0, \ldots, T_k), x) \in A|R) \, \mathsf{P}(C).$$

One has

$$\mathbb{P}(F_{(i,i)}^{k+1,t}((T_0, \ldots, T_k), x) \in A|R)$$

$$= \frac{1}{\mathbb{P}(R)} \int_{\Delta_{k+1,t}} \rho_i(t_0) \prod_{l=1}^{k} \rho_{i_l}(t_l) \mathbf{1}_A(F_{(i,i)}^{k+1,t}((t_0, \ldots, t_k), x))$$

$$\int_{t-(t_0+\ldots+t_k)}^{\infty} \rho_{i_{k+1}}(t_{k+1}) \, dt_{k+1} \, dt_0 \ldots dt_k$$

$$\geq \frac{\tilde{c}}{\mathbb{P}(R)} \lambda^{k+1}(\Delta_{k+1,t})Q(x, A) \geq \frac{\tilde{c}}{\mathbb{P}(R)} \lambda^{k+1}(\Delta_{k+1,t})\xi(A),$$

where

$$\tilde{c} := \inf_{(t_0,\ldots,t_k)\in\Delta_{k+1,t}} \lambda_i \prod_{l=1}^{k} \lambda_{i_l} \exp\left(-\lambda_i t_0 - \sum_{l=1}^{k} \lambda_{i_l}t_l - \lambda_{i_{k+1}}\left(t - \sum_{l=0}^{k} t_l\right)\right) > 0.$$

This proves that (x^*, i) is a Doeblin point for P_t.   □

Corollary 6.31 *If the strong bracket condition holds at an accessible point* $x^* \in M$, *then for all* $s > 0$

$$\mathsf{Inv}(P_s) = \mathsf{Inv}(\{P_t\})$$

and $\mathsf{Inv}(P_s)$ *has at most cardinality one.*

Proof This follows from Proposition 6.3 and Theorem 6.30. $\qquad\qquad \square$

6.5 Stochastic Differential Equations

This section, and the related Sect. 7.6.2, are not self-contained and require some extra knowledge, namely a certain familiarity with stochastic differential equations.

Let $G_0, G_1, \ldots, G_N$ denote smooth vector fields on $M = \mathbb{R}^k$ (or on a k-dimensional manifold). For simplicity, we shall assume here that the vector fields G_i are bounded with bounded first and second derivatives.

We consider the Stratonovich stochastic differential equation on M

$$dX_t = G_0(X_t)\, dt + \sum_{i=1}^{N} G_i(X_t) \circ dB_t^i, \qquad (6.12)$$

where $B = (B_t)_{t \geq 0} = (B_t^1, \ldots, B_t^N)_{t \geq 0}$ is an N-dimensional $\{\mathcal{F}_t\}$-Brownian motion, starting from 0, defined on a probability space $\{\Omega, \mathcal{F}, \mathsf{P}\}$ equipped with a (complete) filtration $\{\mathcal{F}_t\}_{t \geq 0}$.

Equivalently, using the Itô formalism,

$$dX_t = \tilde{G}_0(X_t)\, dt + \sum_{i=1}^{N} G_i(X_t)\, dB_t^i, \qquad (6.13)$$

where

$$\tilde{G}_0(x) = G_0(x) + \frac{1}{2} \sum_{i=1}^{N} DG_i(x) G_i(x).$$

By classical results (see, e.g., [45, Chapter 8]), given $x \in M$ there exists a unique solution, $X^x = (X_t^x)_{t \geq 0}$, to (6.12) with $X_0^x = x$. Furthermore, $X^x \in C(\mathbb{R}_+, M)$ and the mapping $x \mapsto X^x$ is continuous when $C(\mathbb{R}_+, M)$ is equipped with the topology of uniform convergence on compact intervals.

Let $\{P_t\}_{t \geq 0}$ be the family of operators on $B(M)$ defined by

$$P_t f(x) = \mathsf{E}(f(X_t^x)).$$

Then it is also classical that (X_t) is a continuous Feller Markov process with semigroup $\{P_t\}_{t \geq 0}$ (see, e.g., [45] or [59]).

6.5.1   Accessibility

Associated to (6.5) is the deterministic control system

$$\dot{y} = G_0(y) + \sum_{j=1}^{N} u_j(t) G_j(y), \tag{6.14}$$

where $u : [0, \infty) \rightarrow \mathbb{R}^N$ is a *control* function which can be chosen piecewise continuous or piecewise constant.

We let $t \mapsto y(t, x, u)$ denote the solution to (6.14) whose initial condition is x.

Proposition 6.32 *Let* $p, x \in M$. *The following statements are equivalent:*

(i) *For every neighborhood U of p, there exists a control u which can be chosen piecewise continuous or piecewise constant, and $t \geq 0$ such that $y(t, x, u) \in U$;*

(ii) *The point p is accessible from x for $\{P_t\}_{t \geq 0}$.*

Proof This follows from the Stroock–Varadhan support theorem [65], which asserts that the support of the law of X^x equals the closure (in $C(\mathbb{R}_+, M)$) of the set $\{y(\cdot, x, u) : u \text{ piecewise constant}\}$. It is easy to show that the latter also equals the closure of $\{y(\cdot, x, u) : u \text{ piecewise continuous}\}$.   □

6.5.2   Hörmander Conditions

The existence of Doeblin points for the 1-resolvent G or for P_T can be deduced from certain Hörmander conditions that are similar to the bracket conditions introduced in Sects. 6.3.1 and 6.4.2 for PDMPs.

Using the terminology introduced in these sections, we let $\mathcal{L}(G_0, \ldots, G_N)$ denote the Lie algebra generated by $\{G_0, \ldots, G_N\}$, and for all $x \in \mathbb{R}^k$

$$\mathcal{L}(G_0, \ldots, G_N)(x) = \{V(x) : V \in \mathcal{L}(G_0, \ldots, G_N)\}.$$

We define similarly $\mathcal{L}(G_1, \ldots, G_N)$ and $\mathcal{L}(G_1, \ldots, G_N)(x)$.

Given a point $x \in \mathbb{R}^k$ we shall say that x satisfies the *weak Hörmander condition* (respectively the *Hörmander condition*, respectively the *strong Hörmander condition*) if:

(a) [Weak Hörmander condition] $\mathcal{L}(G_0, \ldots, G_N)(x) = \mathbb{R}^k$;
(b) [Hörmander condition] The family

$$\mathcal{L}(G_1, \ldots, G_N)(x) \cup \{[X, Y](x) : X, Y \in \mathcal{L}(G_0, \ldots, G_N)\}$$

spans $\mathbb{R}^k$;
(c) [Strong Hörmander condition] $\mathcal{L}(G_1, \ldots, G_N)(x) = \mathbb{R}^k$.

Clearly $(c) \Rightarrow (b) \Rightarrow (a)$. Observe that in (a) all the vector fields, including the drift G_0, play the same role. In (b) the drift can only appear in a bracket with some "Brownian" vector field. In (c) only the Brownian vector fields appear.

A classical theorem in geometric control theory, originally due to W. L. Chow [15], has the following useful consequence:

Proposition 6.33 *Let $U \subset \mathbb{R}^k$ be a connected open set. Suppose that the strong Hörmander condition holds at every point $x \in U$. Then for all $x, y \in U$ the point y is accessible for $\{P_t\}_{t \geq 0}$ from the point x.*

Proof For $\varepsilon \geq 0$ and $u : [0, \infty) \to \mathbb{R}^N$ a piecewise continuous function, let $t \to y^\varepsilon(t, x, u)$ denote the solution to the ordinary differential equation $\dot{y} = \varepsilon G_0(y) + \sum_{j=1}^{N} u_j(t) G_j(y)$ with initial condition $y^\varepsilon(0, x, u) = x$. Chow's theorem (see, e.g., [15] or [41, Chapter 2, Theorem 3]) asserts that for all $x, y \in U$ there exist a piecewise constant control u with values in $\{-1, 0, 1\}$ and $t \geq 0$ such that $y^0(s, x, u) \in U$ for all $0 \leq s \leq t$ and $y^0(t, x, u) = y$. To shorten notation, set $y^\varepsilon(s) = y^\varepsilon(s, x, u)$ and $y^0(s) = y^0(s, x, u)$. Then, for all $0 \leq s \leq t$,

$$\|y^\varepsilon(s) - y^0(s)\| \leq \varepsilon \|G_0\|_\infty t + K \int_0^s \| y^\varepsilon(r) - y^0(r) \| \, dr,$$

where $K = \sum_{i=1}^{N} \|DG_i\|_\infty$. Thus, by Gronwall's lemma,

$$\|y^\varepsilon(t) - y^0(t)\| \leq e^{Kt} \varepsilon \|G_0\|_\infty t,$$

so that $y^\varepsilon(t) \to y$ as $\varepsilon \to 0$. To conclude observe that $y^\varepsilon(s, x, u) = y(\varepsilon s, x, u^\varepsilon)$ with $u^\varepsilon(s) = u(\frac{s}{\varepsilon})/\varepsilon$. The result then follows from Proposition 6.32. $\square$

The following results, Theorems 6.34 and 6.37, heavily rely on classical papers on hypoelliptic diffusions by Bony [13] and by Ichihara and Kunita [39].

Theorem 6.34 *The following statements hold:*

(i) *Suppose that the weak Hörmander condition holds at $p \in \mathbb{R}^k$. Then p is a Doeblin point for the 1-resolvent G;*

(ii) *Suppose in addition that p is accessible. Then $\{P_t\}_{t\geq 0}$ has at most one invariant probability measure μ. When it exists, μ is absolutely continuous with respect to λ^k (the Lebesgue measure on $\mathbb{R}^k$) and $\operatorname{supp}(\mu) = \Gamma = \overline{\operatorname{Int}(\Gamma)}$, where Γ stands for the accessible set of $\{P_t\}$.*

Proof

(i) Fix $\mathcal{O}$ a neighborhood of p, small enough so that $\mathcal{L}(G_0, \ldots, G_N)(x)$ spans $\mathbb{R}^k$ for all $x \in \mathcal{O}$.

We say that p is *totally degenerate* if $\sum_{i=1}^{N} \|G_i(p)\| = 0$. We distinguish between two cases.

Case 1: p is not totally degenerate.

In this case, there exists a connected open set D containing p, relatively compact, with $\overline{D} \subset \mathcal{O}$ such that:

(a) For every $x \in \overline{D}$, $\sum_{i=1}^{N} \|G_i(x)\| \neq 0$;
(b) For every $x \in \partial D = \overline{D} \setminus D$, there exists a vector u normal to $\overline{D}$ at x such that $\sum_{i=1}^{N} \langle G_i(x), u \rangle^2 > 0$.

Here, by a vector *normal* to $\overline{D}$ at x, we mean that there exists $r > 0$ such that the open ball with center $x + ru$ and radius $r\|u\|$ has empty intersection with $\overline{D}$.

The reason for which such a D exists is the following. We can assume, without loss of generality, that $G_1(p) \neq 0$ and $\frac{G_1(p)}{\|G_1(p)\|} = e_1$, the first vector in the canonical basis of $\mathbb{R}^k$. For $\varepsilon > 0$ small enough, let

$$D = \{x \in \mathbb{R}^k : \|x - p\|_1 < \varepsilon\},$$

where $\|u\|_1 = \sum_{i=1}^{k} |u_i|$. For $x \in \partial D$ let u_x be the vector defined by $u_{x,i} = \frac{x_i - p_i}{|x_i - p_i|}$ if $x_i \neq p_i$ and $u_{x,i} = 1$ otherwise. The vector u_x is normal to ∂D and $\langle e_1, u_x \rangle^2 = 1$. Hence, for ε small enough, $\langle G_1(x), u_x \rangle^2 > 0$ for all $x \in \partial D$ and $G_1(x) \neq 0$ for all $x \in \overline{D}$.

The "formal generator" of the diffusion process (6.12) is the operator L acting on C^2 functions $f : \mathbb{R}^k \to \mathbb{R}$ by the formula

$$Lf = G_0(f) + \frac{1}{2}\sum_{i=1}^{N} G_i^2(f), \tag{6.15}$$

where $G_i(f)(x) = \langle \nabla f(x), G_i(x) \rangle$ and $G_i^2(f) = G_i(G_i(f))$. Under the conditions (a) and (b) above, there exists, by a theorem of Bony [13, Theorem 6.1], a kernel $G_D : \overline{D} \times \overline{D} \to \mathbb{R}_+$, smooth on $D \times D \setminus \{(x, x) : x \in D\}$, such that the following holds:

For every $f \in C_b(\overline{D})$, there exists a unique solution $g \in C_b(\overline{D})$ to the Dirichlet problem

$$\begin{cases} Lg - g = -f \text{ on } D \text{ (in the sense of distributions)} \\ g|_{\partial D} = 0, \end{cases}$$

and $g(x) = G_D f(x) := \int G_D(x, y)f(y)\, dy$. Furthermore, if f is smooth on D so is g.

Note that, by continuity of G_D off the diagonal, there exist disjoint open sets $U, V \subset D$ and $\delta > 0$ such that $p \in U$ and $G_D(x, y) \geq \delta$ for every $(x, y) \in U \times V$.

Let $\tau = \inf\{t > 0 : X_t^x \notin D\}$. For $f \in C_b(D)$ smooth on D, Itô's formula implies that

$$\left(e^{-t \wedge \tau} g(X_{t \wedge \tau}^x) + \int_0^{t \wedge \tau} e^{-s} f(X_s^x)\, ds \right)_{t \geq 0}$$

is a local martingale. Being bounded, it is a uniformly integrable martingale. Thus,

$$\mathsf{E}\left(\int_0^\tau e^{-s} f(X_s^x)\, ds \right) = G_D f(x).$$

It follows that for every $x \in U$ and every Borel set $A \subset \mathbb{R}^k$,

$$G(x, A) \geq G_D \mathbf{1}_A(x) \geq \delta \lambda^k (A \cap V),$$

proving that p is a Doeblin point for G.

Case 2: p is totally degenerate.

Let $\{\Phi_0(t, \cdot)\}$ be the flow induced by G_0. We first assume that $k \geq 2$. We claim that it is possible to choose $t > 0$ small enough to ensure that $\Phi_0(t, p)$ lies in $\mathcal{O}$ and is not totally degenerate. By what precedes, $\Phi_0(t, p)$ is then a Doeblin point for G, and - since it is accessible from p by $\{P_t\}_{t \geq 0}$ -, this makes p a Doeblin point for G (the proof of this latter assertion is easy and left to the reader). To prove the claim, assume to the contrary that $G_i(\Phi_0(t, p)) = 0$ for all $0 < t < \varepsilon$ and $i = 1, \ldots, N$. Then

$$0 = DG_i(\Phi_0(t, p)) \frac{d}{dt} \Phi_0(t, p) = DG_i(\Phi_0(t, p)) G_0(\Phi_0(t, p)) = [G_0, G_i](\Phi_0(t, p)).$$

Similarly $Z(\Phi_0(t, p)) = 0$ for all $Z \in \mathcal{L}(G_0, \ldots, G_N) \setminus \{G_0\}$. This is in contradiction with the assumption that $\mathcal{L}(G_0, \ldots, G_N)$ has rank $k \geq 2$ on $\mathcal{O}$.

Suppose now that $k = 1$. If for some $t > 0$ and $i \in \{1, \ldots, N\}$ $G_i(\Phi_0(t, p)) \neq 0$, the point $\Phi_0(t, p)$ is not totally degenerate, and like previously p is a Doeblin point. If for all $t \geq 0$ and $i \in \{1, \ldots, N\}$

$G_i(\Phi_0(t, p)) = 0$, then for all $x \in \{\Phi_0(t, p) : t > 0\}$ and $f \geq 0$,

$$Gf(x) = \int_0^\infty e^{-t} f(\Phi_0(t, x))dt \geq e^{-1} \int_0^1 f(\Phi_0(t, x))dt = e^{-1} \int_x^{\Phi_0(1,x)} \frac{f(u)}{G_0(u)} du.$$

This easily implies that x, hence p, is Doeblin for G.

(*ii*) Suppose that p is accessible. Then, by Theorem 6.2, G (and hence $\{P_t\}_{t \geq 0}$) has at most one invariant probability measure μ. The minorization $G(x, A) \geq \delta \lambda^k (A \cap V)$ for all $x \in U$ shows that $V \subset \Gamma$. Thus, Γ has nonempty interior and consequently (see Proposition 6.2) $\mathsf{supp}(\mu) = \Gamma$. Also, for every piecewise constant control u, the map $x \mapsto y(t, x, u)$ is a diffeomorphism. The set $\cup_{t,u} y(t, V, u)$, with the union taken over all $t \geq 0$ and u piecewise constant, is then an open set dense in Γ. It remains to prove that $\mu \ll \lambda^k$. Let $C(\mathbb{R}_+, \mathbb{R}^N)$ be the Wiener space equipped with its Borel σ-field and the Wiener measure $W(dw)$ (i.e., the law of $B = (B_t^1, \ldots, B_t^N)_{t \geq 0}$) and let $\Theta = \mathbb{R}_+ \times C(\mathbb{R}_+, \mathbb{R}^N)$ be equipped with the product measure $m(dt\,dw) = e^{-t} dt\, W(dw)$. Then, for all $f \in B(M)$,

$$Gf(x) = \int_\Theta f(F_\theta(x))\, m(d\theta),$$

where $F_{(t,w)}(x) = X_t^x(w)$. Now, for almost all w and all $t \geq 0$, the map $x \mapsto X_t^x(w) = F_{(t,w)}(x)$ is a diffeomorphism (see, e.g., [40, Chapter V] or Kunita [44]). We are then in the situation already considered in Theorem 6.9 and the proof of Theorem 6.9 applies verbatim.

□

Remark 6.35 Suppose that $\Gamma \neq \emptyset$ and that **all** the points in Γ satisfy the weak Hörmander condition. Then the density of μ (when μ exists) is C^∞. Indeed, let U be a neighborhood of Γ such that all the points in U satisfy the weak Hörmander condition. By Hörmander's theorem [38], L and L^* are *hypoelliptic* operators in U, meaning that for every distribution f on U, $Lf \in C^\infty(U) \Rightarrow f \in C^\infty(U)$. If $f = \frac{d\mu}{d\lambda^k}$, $L^* f = 0$ so that f is smooth.

Remark 6.36 Suppose that all the points in M satisfy the strong Hörmander condition (and, in case M is a manifold, M is connected). Then $\Gamma = M$ and the density of μ (when μ exists) is positive everywhere. The first statement follows from Proposition 6.33 and the second from Bony's maximum principle [13, Corollaire 3.1] applied to L^*.

Theorem 6.37 *Let $p \in \mathbb{R}^k$. Suppose that the Hörmander condition holds at p. Then p is a Doeblin point for some P_t with $t > 0$. If furthermore p is accessible, then for all $s > 0$*

$$\mathsf{Inv}(P_s) = \mathsf{Inv}(\{P_t\})$$

and $\mathsf{Inv}(P_s)$ has at most cardinality one.

Proof Let D be a neighborhood of p at which the Hörmander condition holds. Then the law of (X_t) killed at D (see Ichihara and Kunita [39]) has a density $q_t(x, y)$ which is C^∞ in $t > 0, x, y \in D$. Thus, $q_t(p, q) > 0$ for some $t > 0$ and $q \in D$. This makes p a Doeblin point for P_t. The second statement follows from Proposition 6.3. $\qquad\Box$

Notes

The material on random switching between vector fields in Sect. 6.3 is based on [8] and [5]. More background on the weak and strong bracket conditions with an emphasis on how they relate to controllability is provided in [66]. Proposition 6.3 and the material in Sect. 6.5 are based on [6] and [10]. The first proof that, under a weak Hörmander condition at an accessible point, an SDE has at most one invariant probability measure goes back to Arnold and Kliemann [2]. The proof given here (of Theorem 6.34) is based on the notes [6] and differs from the proof in [2].

Chapter 7
Harris and Positive Recurrence

Positive recurrent chains are uniquely ergodic chains for which the Birkhoff ergodic theorem (i.e., the strong law of large numbers) holds true for **every** initial condition. Harris recurrent chains are chains which satisfy a weaker form of recurrence (defined below). It turns out that Harris recurrent chains possessing an invariant probability measure are positive recurrent. Several criteria ensuring Harris and positive recurrence are given in this chapter. These are applied in the final section to Feller chains, piecewise deterministic Markov processes, and stochastic differential equations.

7.1 Stability and Positive Recurrence

Let (X_n) denote a Markov chain (defined on $(\Omega, \mathcal{F}, \mathbb{F}, \mathsf{P})$) on M with kernel P. Recall that we let

$$\nu_n = \frac{1}{n} \sum_{i=1}^{n} \delta_{X_k}$$

denote its *empirical occupation measure*.

If there exists $\pi \in \mathcal{P}(M)$ such that, for all $x \in M$ and every bounded **continuous** (respectively **measurable**) function $f : M \to \mathbb{R}$,

$$\mathbb{P}_x(\lim_{n \to \infty} \nu_n f = \pi f) = 1,$$

the kernel P (or the chain (X_n)) is called *stable*, respectively *positive recurrent*.

If P is stable, then it is clearly uniquely ergodic with invariant probability measure π, where π is the probability measure appearing in the definition.

The following partial converse follows from Theorem 4.20.

© The Author(s), under exclusive license to Springer Nature Switzerland AG 2022
M. Benaïm, T. Hurth, *Markov Chains on Metric Spaces*, Universitext,
https://doi.org/10.1007/978-3-031-11822-7_7

Proposition 7.1 *Suppose that P is Feller, uniquely ergodic, and that for all $x \in M$, $\{v_n\}$ is $\mathbb{P}_x$-almost surely tight. Then P is stable.*

Remark 7.2 A stable Feller Markov chain is not necessarily positive recurrent. For instance, let $X_n \in [-2, 2]$ be recursively defined as

$$X_{n+1} = \frac{1}{2}X_n + \xi_{n+1},$$

where (ξ_n) are independent uniformly distributed random variables taking values in $\{-1, 1\}$. Then (X_n) is Feller and uniquely ergodic (see, e.g., Exercise 3.9 or Theorem 4.31), hence stable. It is not hard to prove that π, its stationary distribution, is the uniform distribution over $[-2, 2]$. On the other hand, for $X_0 = 0$, $X_n \in D = \{\sum_{k=0}^{m} 2^{-k}\theta_k : \theta_k \in \{-1, 1\}, m \in \mathbb{N}\}$ so that $v_n(D) = 1$ while $\pi(D) = 0$.

Another example (borrowed from [22]) is the following. Let P be the kernel on $[0, \infty)$ defined by $P(0, 0) = 1$ and, for $x > 0$, $P(x, 0) = 1 - P(x, x/2) = 2^{-x}$. This kernel is δ_0-irreducible, Feller, and admits δ_0 as the unique invariant probability measure. It is stable (since $X_n \leq \frac{X_0}{2^n}$) but is not positive recurrent because the probability that X_n never touches 0 is positive.

Proposition 7.3 *Suppose that P is strong Feller and stable. Then P is positive recurrent.*

Proof If P is strong Feller, then, for every bounded measurable f, Pf is continuous so that $v_n(Pf) - \pi(Pf) \to 0$, $\mathbb{P}_x$-almost surely. By invariance of π, $\pi(Pf) = \pi f$ and, as shown in the proof of Theorem 4.20, $v_n(Pf) - v_n f \to 0$, $\mathbb{P}_x$-almost surely.
□

Remark 7.4 A Feller (even strong Feller) uniquely ergodic kernel on a noncompact space is not necessarily stable. For instance, let P be the kernel on $\mathbb{N}$ defined as $P(0, 0) = 1$ and, for $n \geq 1$, $P(n, n - 1) = 1 - p$, $P(n, n + 1) = p$ with $1 > p > 1/2$. Then δ_0 is the unique invariant probability measure of this Markov chain but the chain is not stable since $\mathbb{P}_x(X_n \to \infty) > 0$ for all $x > 0$. Another (similar) example on $\mathbb{R}^d$ is given by the deterministic linear dynamical system $X_{n+1} = aX_n$ with $a > 1$.

Exercise 7.5 Let (X_n) be the deterministic system on $S^1 = \mathbb{R}/\mathbb{Z}$ defined by $X_{n+1} = (X_n + \alpha) \mod 1$, where $\alpha \in \mathbb{R} \setminus \mathbb{Q}$. Show that (X_n) is stable but not positive recurrent.

7.2 Harris Recurrence

The chain (X_n) is called *Harris recurrent* if there exists a nonzero measure ξ such that for every Borel set $A \subset M$ and every $x \in M$,

$$\xi(A) > 0 \Rightarrow \mathbb{P}_x\left(\limsup_{n\to\infty} \mathbf{1}_A(X_n) = 1 \right) = 1.$$

Note that a Harris recurrent chain is ξ-irreducible. The converse is false as shown by the following example.

Example 7.6 Let P be the Markov transition matrix on $M = \mathbb{N}$ defined by $P(i, i + 1) = p_i$ and $P(i, 0) = 1 - p_i$, where $p_0 = 0$, $p_i > 0$ for $i \geq 1$, and $\prod_{i\geq 1} p_i > 0$. Then the associated chain is δ_0-irreducible but not Harris recurrent.

Recall that a *harmonic* function is a measurable function $h : M \to \mathbb{R}$ such that

$$Ph = h.$$

Theorem 7.7 *Suppose that (X_n) is Harris recurrent. Then every bounded harmonic function is constant.*

Proof Let h be bounded and harmonic. Let (X_n^x) denote the chain having P as Markov kernel and initial condition $X_0^x = x$. Then $Y_n = h(X_n^x)$ is a bounded (in particular uniformly integrable) martingale. Hence, by Doob's convergence theorem (Theorem A.7 in the appendix), $\lim_{n\to\infty} Y_n = Y_\infty$ exists almost surely and $\mathbb{E}(Y_\infty | \mathcal{F}_n) = Y_n$. Given $a \in \mathbb{R}$, let $\{h \geq a\}$ (respectively $\{h \leq a\}$, $\{h = a\}$) be the set of $u \in M$ such that $h(u) \geq a$ (respectively $\leq, =$). If $\xi(\{h \geq a\}) > 0$, then (X_n^x) enters $\{h \geq a\}$ infinitely often. Thus $Y_\infty \geq a$ so that $Y_n = \mathbb{E}(Y_\infty | \mathcal{F}_n) \geq a$. In particular, $h(x) = Y_0 \geq a$. Similarly if $\xi(\{h \leq a\}) > 0$ then $h(x) \leq a$. Let now a be such that $\{h = a\} \neq \emptyset$. Then $\xi(\{h \neq a\}) = \xi(\cup_{n\in\mathbb{N}}\{a - (n + 1)^{-1} \leq h \leq a + (n + 1)^{-1}\}^c) = 0$. This proves that $h = a$. $\square$

Positive recurrence and Harris recurrence are intimately linked as shown by the next important theorem.

Theorem 7.8 *The following assertion are equivalent:*

(a) *P is Harris recurrent and $\mathsf{Inv}(P) \neq \emptyset$;*
(b) *P is positive recurrent;*
(c) *There exists $\pi \in \mathsf{Inv}(P)$ such that for all $f \in L^1(\pi)$ and every initial distribution μ,*

$$\mathbb{P}_\mu(\lim_{n\to\infty} \nu_n(f) = \pi(f)) = 1.$$

Proof $(c) \Rightarrow (b) \Rightarrow (a)$ is immediate. Conversely, if P is Harris recurrent with an invariant probability measure π, then P is uniquely ergodic. Let $f \in L^1(\pi)$, $\mathcal{A} = \{\omega \in M^{\mathbb{N}} : \lim_{n \to \infty} \frac{1}{n} \sum_{k=1}^{n} f \circ \theta^k(\omega) = \pi f\}$, and $g(x) = \mathbb{P}_x(\mathcal{A})$. By the ergodic theorem, $g(x) = 1$, π-almost surely. We now claim that g is harmonic, which with Theorem 7.7 proves the result. To prove the claim we use the invariance of $\mathcal{A}$ under θ and the Markov property:

$$g(x) = \mathbb{E}_x(\mathbf{1}_{\mathcal{A}}) = \mathbb{E}_x(\mathbf{1}_{\mathcal{A}} \circ \theta) = \mathbb{E}_x(\mathbb{E}_x(\mathbf{1}_{\mathcal{A}} \circ \theta)|\mathcal{F}_1)) = \mathbb{E}_x(g(X_1)) = Pg(x).$$

$\square$

Theorem 7.9 *Suppose P is strong Feller and uniquely ergodic with an invariant probability measure π having full support. Then the equivalent conditions of Theorem 7.8 hold true.*

Proof Let $f \in L^1(\pi)$ and let g be defined as in the proof of Theorem 7.8. We have seen that g is harmonic. Since P is strong Feller, g is continuous and, by the ergodic theorem, $g(x) = 1$ for π-almost all x. The set $\{x \in M : g(x) = 1\}$ is then a closed set containing the support of π. Since π has full support, $g = 1$ and P is positive recurrent. $\square$

Corollary 7.10 *Suppose P is strong Feller with an invariant probability measure π having full support. If M is connected, then the equivalent conditions of Theorem 7.8 hold true.*

Proof This follows from Theorem 7.9 and Proposition 5.18. $\square$

7.2.1   Petite Sets and Harris Recurrence

A convenient and practical way to ensure that a chain is Harris recurrent is to exhibit a *recurrent petite set*.

Given a Borel set $C \subset M$ we say that $x \in M$ *leads almost surely to* C if $\mathbb{P}_x(\tau_C < \infty) = 1$, where

$$\tau_C = \min\{k \geq 1 : X_k \in C\}.$$

We say that C is *recurrent* if **every** $x \in M$ leads almost surely to C.

For further reference, we define the successive return times in C recursively by

$$\tau_C^{(n+1)} = \min\{k > \tau_C^{(n)} : X_k \in C\}$$

with $\tau_C^{(0)} = 0$.

Proposition 7.11 *Let* $C \subset M$ *be a recurrent petite set. Then* (X_n) *is Harris recurrent.*

Proof It easily follows from the definition of a petite set (see Sect. 6.1) that for all $x \in C$ and A Borel, $\mathbb{P}_x(\tau_A < \infty) \geq \xi(A)$. Thus, using the strong Markov property, for all $x \in M$,

$$\mathbb{P}_x(\tau_A < \infty) \geq \mathbb{P}_x(\exists k \geq \tau_C : X_k \in A) = \mathbb{E}_x(\mathbb{P}_{X_{\tau_C}}(\tau_A < \infty)) \geq \xi(A).$$

Therefore, by the Markov property, for all $n \in \mathbb{N}$

$$\mathsf{P}(\tau_A < \infty | \mathcal{F}_n) = \mathbb{P}_{X_n}(\tau_A < \infty) \geq \xi(A).$$

The first term of this inequality converges to $\mathbf{1}_{\tau_A < \infty}$ (see Theorem A.7 in the appendix). Thus $\mathbb{P}_x(\tau_A < \infty) = 1$ for all x whenever $\xi(A) > 0$. By the strong Markov property, this implies that $X_n \in A$ infinitely often. □

7.3 Recurrence Criteria and Lyapunov Functions

We discuss here simple useful criteria, based on Lyapunov functions, ensuring that a set is recurrent. They also provide moment estimates of the return times. Conditions (*a*) and (*b*) of the next result are folklore (see the notes at the end of the chapter). We learned condition (*a'*) from Philippe Robert (see [60], Proposition 8 in Chapter 8).

Proposition 7.12 *Let* $V : M \to [1, \infty)$ *be a measurable map and* $C \subset M$ *a Borel set. Assume that for all* $x \in C$, $PV(x) < \infty$ *and that one of the three following conditions holds:*

(a) $PV - V \leq -1$ *on* $M \setminus C$;
(a') *Condition* (a) *and* $\sup_{x \in M} \mathbb{E}_x(|V(X_1) - V(x)|^p) < \infty$ *for some* $p > 1$;
(b) $PV - V \leq -\lambda V$ *on* $M \setminus C$ *for some* $1 > \lambda > 0$.

Then for all $x \in M$

(i) $\mathbb{E}_x(\tau_C) \leq PV(x) + 1$ *under condition* (a);
(ii) $\mathbb{E}_x(\tau_C^p) \leq c(1 + V^p(x))$ *for some constant* $c > 0$ *under condition* (a');
(iii) $\mathbb{E}_x(e^{\lambda \tau_C}) \leq \mathbb{E}_x(e^{-\log(1-\lambda)\tau_C}) \leq \frac{1}{1-\lambda} PV(x)$ *under condition* (b).

In particular, C *is a recurrent set.*

Proof Let $V_n = V(X_{n \wedge \tau_C}) + (n \wedge \tau_C)$. Then $(V_n)_{n \geq 1}$ is a supermartingale. Indeed, for all $n \geq 1$,

$$\mathsf{E}(V_{n+1} - V_n | \mathcal{F}_n) = \mathsf{E}(V_{n+1} - V_n | \mathcal{F}_n)\mathbf{1}_{\tau_C > n} = (PV(X_n) - V(X_n))\mathbf{1}_{\tau_C > n} \leq 0.$$

Thus $\mathbb{E}_x(n \wedge \tau_C) \le \mathbb{E}_x(V_n) \le \mathbb{E}_x(V_1) = PV(x) + 1$. This proves the first assertion. The proof of assertion (iii) is similar: Set $V_n = \frac{V(X_{n \wedge \tau_C})}{(1-\lambda)^{n \wedge \tau_C}}$. Then $(V_n)_{n \ge 1}$ is a supermartingale. Thus

$$\mathbb{E}_x(e^{-\log(1-\lambda)n \wedge \tau_C}) \le \mathbb{E}_x(V_n) \le \mathbb{E}_x(V_1) = \frac{PV(x)}{1-\lambda}.$$

We now prove assertion (ii), following Robert ([60], Proposition 8, Chapter 8). We claim that for all $x > -1$

$$(1+x)^p \le 1 + px + C_p\, r(x), \tag{7.1}$$

where

$$r(x) = x^2(1+|x|)^{p-2} \text{ and } C_p = \frac{p(p-1)}{4}$$

for $p \ge 2$; and

$$r(x) = |x|^p \text{ and } C_p = 1$$

for $1 < p < 2$. Indeed, by the Taylor–Lagrange formula, for all $x > -1$,

$$(1+x)^p = 1 + px + \frac{p(p-1)}{2}x^2 R(x)$$

with $R(x) = \int_0^1 (1-s)(1+sx)^{p-2}\, ds$. Thus $|R(x)| \le \frac{1}{2}(1+|x|)^{p-2}$ for $p \ge 2$. For $1 < p < 2$ and $x > 0$, $|R(x)| \le \int_0^1 (1-s)s^{p-2}x^{p-2}\, ds = \frac{1}{p(p-1)}x^{p-2}$, while for $1 < p < 2$ and $-1 < x \le 0$ one has $|R(x)| \le 1$ (because $s \in [0,1] \mapsto (1-s)(1+sx)^{p-2}$ is decreasing, hence bounded above by 1). This proves the claim.
    Now set

$$Z_n = 1 + \varepsilon\left(V(X_n) + \frac{n}{2}\right),$$

where $\varepsilon > 0$ and

$$\Delta_{n+1} = V(X_{n+1}) - V(X_n) + \frac{1}{2}.$$

Then

$$Z_{n+1}^p = Z_n^p\left(1 + \frac{\varepsilon \Delta_{n+1}}{Z_n}\right)^p,$$

so that by (7.1) and condition (a),

$$\mathbb{E}_x(Z_{n+1}^p|\mathcal{F}_n) \leq Z_n^p\left[1 - \frac{p\varepsilon}{2Z_n} + C_p\mathbb{E}_x(r(\frac{\varepsilon\Delta_{n+1}}{Z_n})|\mathcal{F}_n)\right]$$

on the event $\tau_C > n$. Now, it is easy to check that $r(\frac{\varepsilon\Delta_{n+1}}{Z_n}) \leq \frac{\varepsilon^2}{Z_n}(1 + |\Delta_{n+1}|)^p$ for $p \geq 2$, and $r(\frac{\varepsilon\Delta_{n+1}}{Z_n}) \leq \varepsilon^p\frac{|\Delta_{n+1}|^p}{Z_n}$ for $1 < p < 2$. Thus, for $\varepsilon > 0$ small enough, conditions (a) and (a') make $(Z_{n\wedge\tau_C}^p)$ a supermartingale. We can then conclude as in the proof of (i). $\square$

Remark 7.13 If V is a Lyapunov function in the sense that $PV \leq \rho V + \kappa$ with $0 \leq \rho < 1$ and $\kappa \geq 0$, then the assumptions of Proposition 7.12 (b) hold true with $0 < \lambda < 1 - \rho$ and $C = \{x \in M : V(x) \leq \frac{\kappa+1}{1-\rho-\lambda}\}$. Compare to Proposition 4.23.

The next proposition extends assertion (iii) of Proposition 7.12 and gives an alternative condition (to conditions (a), (a')) to control the moments of τ_C. The proof is based on a beautiful argument used in section 4.1 of Hairer's notes [31].

Proposition 7.14 *Let $V : M \to [1, \infty)$ be a measurable map and $C \subset M$ a Borel set. Let $\varphi : [0, \infty) \mapsto \mathbb{R}_+^*$ be a concave C^1-function and let $h : [1, \infty) \to [0, \infty)$ be the map defined by*

$$h(x) = \int_1^x \frac{ds}{\varphi(s)}.$$

Assume that for all $x \in C$, $PV(x) < \infty$ and that for all $x \in M \setminus C$

$$PV(x) - V(x) \leq -\varphi(V(x)).$$

Then, for all $x \in M \setminus C$,

$$\mathbb{E}_x(h^{-1}(\tau_C)) \leq V(x)$$

and, for all $x \in C$,

$$\mathbb{E}_x(h^{-1}(\tau_C)) \leq h^{-1}(h(PV(x)) + 1).$$

Proof First observe that $\varphi' \geq 0$ (for otherwise by concavity φ could not be > 0). For $x \geq 1$ and $t \geq 0$ set $H(t, x) = h^{-1}(h(x) + t)$. It is readily seen that

$$\frac{\partial H}{\partial t}(t, x) = \varphi(H(t, x)) = \varphi(x)\frac{\partial H}{\partial x}(t, x). \tag{7.2}$$

Thus

$$\frac{\partial^2 H}{\partial x^2}(t, x) = \frac{(\varphi'(H(t, x)) - \varphi'(x))\varphi(H(t, x))}{\varphi(x)^2} \le 0. \tag{7.3}$$

In particular, H is convex in t and concave in x.

It follows that for all $n \ge 0$

$$H(n + 1, V(X_{n+1})) - H(n, V(X_n)) =$$

$$H(n + 1, V(X_{n+1})) - H(n + 1, V(X_n)) + H(n + 1, V(X_n)) - H(n, V(X_n))$$

$$\le \frac{\partial H}{\partial x}(n + 1, V(X_n))(V(X_{n+1}) - V(X_n)) + \frac{\partial H}{\partial t}(n + 1, V(X_n)).$$

Therefore, on the event $\{X_n \notin C\}$,

$$\mathbb{E}(H(n + 1, V(X_{n+1})) - H(n, V(X_n))|\mathcal{F}_n)$$

$$\le -\varphi(V(X_n))\frac{\partial H}{\partial x}(n + 1, V(X_n)) + \frac{\partial H}{\partial t}(n + 1, V(X_n)) \le 0.$$

Here the first inequality follows from the hypotheses on V and the second one from Eq. (7.2). This makes the process $(H(n \wedge \tau_C, V(X_{n \wedge \tau_C})))_{n \ge 1}$ a supermartingale. Thus

$$\mathbb{E}_x(h^{-1}(n \wedge \tau_C)) \le \mathbb{E}_x(H(n \wedge \tau_C, V(X_{n \wedge \tau_C}))) \le \mathbb{E}_x(H(1, V(X_1))) \le H(1, PV(x)),$$

where the last inequality follows from concavity of H in x and Jensen's inequality. In the case $x \in M \setminus C$, by monotonicity and concavity of h,

$$h(PV(x)) \le h(V(x) - \varphi(V(x))) \le h(V(x)) - h'(V(x))\varphi(V(x)) = h(V(x)) - 1.$$

Thus $H(1, PV(x)) \le V(x)$. This proves the result. $\square$

7.4   Subsets of Recurrent Sets

Let $C \subset M$ be a recurrent set for the chain (X_n) (for instance the sublevel set $\{V \le R\}$ of a Lyapunov function) and $U \subset C$ a measurable smaller subset (for instance the neighborhood of a Doeblin point). It is often desirable to deduce recurrence properties of U from recurrence properties of C. This short section discusses two such results.

The *induced chain* on C is the process $(Y_n)_{n\geq 1}$ defined as

$$Y_n = X_{\tau_C^{(n)}}.$$

Exercise 7.15 Verify that $(Y_n)_{n\geq 1}$ is a Markov chain on C.

Proposition 7.16 *Let $C \subset M$ be a nonempty recurrent set and $U \subset C$ a measurable subset. Suppose that there exist $k \geq 1$ and $0 < \varepsilon \leq 1$ such that for all $x \in C$*

$$\mathbb{P}_x(\exists i \in \{1, \ldots, k\} : Y_i \in U) \geq \varepsilon,$$

where (Y_n) is the induced chain on C. Then

  (i) *U is recurrent;*
 (ii) *If $\sup_{x\in C} \mathbb{E}_x(\tau_C^p) < \infty$ for some $p \geq 1$, then*

$$\sup_{x\in C} \mathbb{E}_x(\tau_U^p) < \infty;$$

(iii) *If $\sup_{x\in C} \mathbb{E}_x(e^{\lambda_0\tau_C}) < \infty$ for some $\lambda_0 > 0$, then*

$$\sup_{x\in C} \mathbb{E}_x(e^{\lambda\tau_U}) < \infty$$

  for some $0 < \lambda \leq \lambda_0$.

Proof For all $x \in M$, $\mathbb{P}_x$-almost surely,

$$\mathbf{1}_{\tau_U < \infty} = \lim_{n\to\infty} \mathbb{P}_x(\tau_U < \infty | \mathcal{F}_{\tau_C^{(n)}}) = \lim_{n\to\infty} \mathbb{P}_{Y_n}(\tau_U < \infty) \geq \varepsilon.$$

Here the first equality follows from the martingale convergence theorem A.7 and the second from the strong Markov property. This proves that U is recurrent.

Let $\sigma_U = \min\{n \geq 1 : Y_n \in U\}$. The proofs of assertions (ii) and (iii) now follow from the identity $\tau_U = \tau_C^{(\sigma_U)}$, exactly as in the proof of Proposition 2.18 (i), (ii). The verification is an easy exercise left to the reader. $\square$

When P is Feller, the existence of a compact recurrent set C makes every accessible open set U recurrent. More precisely,

Proposition 7.17 *Suppose that P is Feller. Let $C \subset M$ be a nonempty compact set, $x^* \in M$ an accessible point from C (i.e., $x^* \in \Gamma_C$) and U a neighborhood of x^*.*

 (i) *If C is recurrent, so is U;*
(ii) *If $U \subset C$ and $\sup_{x\in C} \mathbb{E}_x(\tau_C^p) < \infty$ for some $p \geq 1$, then*

$$\sup_{x\in C} \mathbb{E}_x(\tau_U^p) < \infty;$$

(iii) *If $U \subset C$ and $\sup_{x \in C} \mathbb{E}_x(e^{\lambda_0 \tau_C}) < \infty$ for some $\lambda_0 > 0$, then*

$$\sup_{x \in C} \mathbb{E}_x(e^{\lambda \tau_U}) < \infty$$

for some $0 < \lambda \leq \lambda_0$.

Proof For $\varepsilon > 0$ and $i \in \mathbb{N}^*$ let $O(\varepsilon, i) = \{x \in M : P^i(x, U) > \varepsilon\}$. By Feller continuity and the Portmanteau theorem 4.1, $O(\varepsilon, i)$ is an open set. By accessibility of x^*, the family $\{O(\varepsilon, i) : \varepsilon > 0, i \in \mathbb{N}^*\}$ covers C. Thus, by compactness, there exist $\varepsilon > 0$ and a finite set $I \subset \mathbb{N}$ such that $C \subset \cup_{i \in I} O(\varepsilon, i)$. This shows that, for all $x \in C$,

$$\mathbb{P}_x(\tau_U \leq k) \geq \varepsilon \tag{7.4}$$

with k the largest element of I. Assertions (ii) and (iii) then follow from Proposition 7.16 because, for all $x \in C$,

$$\mathbb{P}_x(\exists i \in \{1, \ldots, k\} : Y_i \in U) \geq \mathbb{P}_x(\tau_U \leq k) \geq \varepsilon.$$

The proof of the first assertion is similar to the proof of the first assertion in Proposition 7.16. Namely, for all $x \in M$, $\mathbb{P}_x$-almost surely,

$$\mathbf{1}_{\tau_U < \infty} = \lim_{n \to \infty} \mathbb{P}_x(\tau_U < \infty | \mathcal{F}_{\tau_C^{(n)}}) = \lim_{n \to \infty} \mathbb{P}_{Y_n}(\tau_U < \infty) \geq \varepsilon.$$

Thus $\mathbb{P}_x(\tau_U < \infty) = 1$. □

7.5   Petite Sets and Positive Recurrence

We have seen (Proposition 7.11) that the existence of a recurrent petite set for a Markov chain implies Harris recurrence of the chain. If, in addition, the return times to the set are bounded in L^1, then the chain is positive recurrent.

Theorem 7.18 *Let $C \subset M$ be a recurrent petite set such that*

$$\sup_{x \in C} \mathbb{E}_x(\tau_C) < \infty.$$

Then the equivalent conditions of Theorem 7.8 hold true.

Before proving this theorem, we start with a proposition relating the recurrence properties of the chain (X_n) and the sampled chain $Y_n := X_{T_n}$, where

$$T_n := \Delta_1 + \ldots + \Delta_n$$

for $n \geq 1$, $T_0 := 0$, and $(\Delta_i)_{i \geq 1}$ is a sequence of i.i.d. random variables taking on values in $\mathbb{N}$.

Recall that in the particular case where Δ_i has a geometric distribution with parameter a (i.e., $P(\Delta_i = n) = a^n(1 - a)$ for all $n \in \mathbb{N}$), then (Y_n) has kernel R_a.

The *hazard rate* of Δ_i is the sequence

$$\lambda(n) = P(\Delta_i = n | \Delta_i \geq n) = \frac{P(\Delta_i = n)}{P(\Delta_i \geq n)}, \quad n \in \mathbb{N}.$$

For a geometric distribution with parameter a, the hazard rate is constant and equals $1 - a$.

Exercise 7.19 Suppose Δ_i has a negative binomial distribution with parameters (a, m) (see Exercise 5.2 (ii)). Prove that $\lambda(n)$ is nondecreasing and converges to $1 - a$. In particular,

$$\inf_{n \in \mathbb{N}} \lambda(n) = \lambda(0) = (1 - a)^m.$$

The next result is an easy consequence of the memoryless property when Δ_i has a geometric distribution (prove it as an exercise) and this is exactly what we'll need for the proof of Theorem 7.18. It is however interesting to point out that it remains valid under the weaker assumption that the hazard rate of Δ_i is bounded below. Tom Mountford helped us with the proof of this proposition and suggested the minorization condition on the hazard rate.

Proposition 7.20 *Let (Δ_n), (T_n) be as above, i.e., (Δ_n) is an i.i.d. sequence of $\mathbb{N}$-valued random variables and $T_n := \Delta_1 + \ldots + \Delta_n$. Assume that there is $\alpha \in (0, 1)$ such that*

$$\inf_{n \in \mathbb{N}} \lambda(n) \geq 1 - \alpha > 0.$$

Let $\mathcal{N} = \{n_1 < n_2 < \ldots < n_k < \ldots\} \subset \mathbb{N}$ be an infinite set of integers and

$$\tau_{\mathcal{N}} := \min\{n \geq 1 : T_n \in \mathcal{N}\}.$$

Then

  (i) $P(\tau_{\mathcal{N}} < \infty) = 1$;
 (ii) $P(T_{\tau_{\mathcal{N}}} > n_i) \leq \alpha^i$ *for all* $i \geq 1$;
(iii) $E(\Delta_1)E(\tau_{\mathcal{N}}) \leq n_1 + \sum_{i \geq 1}(n_{i+1} - n_i)\alpha^i$;
 (iv) *If $\lambda(n) = 1 - \alpha$ for all $n \in \mathbb{N}$ (meaning that Δ_i has a geometric distribution with parameter α), inequalities (ii) and (iii) are equalities.*

Proof

(i) For $n \geq 1$, let $\mathcal{F}_n := \sigma(\Delta_1, \ldots, \Delta_n)$ and $v(n) := P(\exists i \geq 0 : T_i = n)$. We claim that $v(n) \geq 1 - \alpha$ for all $n \geq 1$. One has

$$v(n) = E(P(\exists i \geq 0 : T_i = n | \mathcal{F}_1))$$
$$= v(n)P(\Delta_1 = 0) + E(v(n - \Delta_1)1_{0 < \Delta_1 < n}) + P(\Delta_1 = n).$$

Thus, $v(1) = \lambda(1) \geq 1 - \alpha$. Suppose now that $v(i) \geq 1 - \alpha$ for $i = 1, \ldots, n - 1$. Then

$$v(n)P(\Delta_1 > 0) \geq (1-\alpha)P(0 < \Delta_1 < n) + P(\Delta_1 = n) \geq (1-\alpha)P(\Delta_1 > 0).$$

This proves the claim by induction. It follows from what precedes that $P(\tau_\mathcal{N} < \infty | \mathcal{F}_n) \geq 1 - (1 - \alpha)^n$, so that P-almost surely

$$1_{\tau_\mathcal{N} < \infty} = \lim_{n \to \infty} P(\tau_\mathcal{N} < \infty | \mathcal{F}_n) = 1.$$

(ii) For $k \geq 1$, let $S_k := \min\{i \geq 0 : n_k \leq T_i < n_{k+1}\} \in \mathbb{N} \cup \{\infty\}$. Then

$$P(T_{\tau_\mathcal{N}} > n_{k+1}) = P(T_{\tau_\mathcal{N}} > n_{k+1}; S_k < \infty) + P(T_{\tau_\mathcal{N}} > n_{k+1}; S_k = \infty).$$

Using the strong Markov property,

$$P(T_{\tau_\mathcal{N}} > n_{k+1}; S_k < \infty) = E(P(T_{\tau_\mathcal{N}} > n_{k+1} | \mathcal{F}_{S_k})1_{\{S_k < \infty\}})$$

$$= E((1 - v(n_{k+1} - T_{S_k}))1_{\{T_{\tau_\mathcal{N}} > n_k\}}1_{\{S_k < \infty\}}) \leq \alpha P(T_{\tau_\mathcal{N}} > n_k; S_k < \infty).$$

On the other hand,

$$P(T_{\tau_\mathcal{N}} > n_{k+1}; S_k = \infty)$$
$$= \sum_{i \geq 0} P(\{T_0, T_1, \ldots, T_i\} \cap \{n_1, \ldots, n_k\} = \emptyset; T_i < n_k; T_{i+1} > n_{k+1}),$$

and

$$P(\{T_0, T_1, \ldots, T_i\} \cap \{n_1, \ldots, n_k\} = \emptyset; T_i < n_k; T_{i+1} > n_{k+1} | \mathcal{F}_i)$$

$$= 1_{\{T_0, T_1, \ldots, T_i\} \cap \{n_1, \ldots, n_k\}}1_{T_i < n_k}P(\Delta_{i+1} > n_{k+1} - T_i | \mathcal{F}_i)$$

$$\leq \alpha 1_{\{T_0, T_1, \ldots, T_i\} \cap \{n_1, \ldots, n_k\}}1_{T_i < n_k}P(\Delta_{i+1} \geq n_{k+1} - T_i | \mathcal{F}_i)$$

by the assumption on the hazard rate of (Δ_i). Therefore,

$$P(T_{\tau_{\mathcal{N}}} > n_{k+1}; S_k = \infty)$$

$$\leq \alpha \sum_{i \geq 0} E(\mathbf{1}_{\{T_0, T_1, \ldots, T_i\} \cap \{n_1, \ldots, n_k\}} \mathbf{1}_{T_i < n_k} P(\Delta_{i+1} \geq n_{k+1} - T_i | \mathcal{F}_i))$$

$$= \alpha P(T_{\tau_{\mathcal{N}}} > n_k; S_k = \infty).$$

Finally we have shown that

$$P(T_{\tau_{\mathcal{N}}} > n_{k+1}) \leq \alpha P(T_{\tau_{\mathcal{N}}} > n_k).$$

(iii) Let $M_n := T_n - E(T_n) = T_n - nm$, where $m := E(\Delta_i)$. Then (M_n) is an $(\mathcal{F}_n)$-martingale with zero mean. Thus, by part (ii) of Theorem A.4,

$$E(M_{n \wedge \tau_{\mathcal{N}}}) = 0 = E(T_{\tau_{\mathcal{N}} \wedge n}) - mE(\tau_{\mathcal{N}} \wedge n),$$

and, by monotone convergence,

$$mE(\tau_{\mathcal{N}}) = E(T_{\tau_{\mathcal{N}}}) = \sum_{k \geq 1} n_k P(T_{\tau_{\mathcal{N}}} = n_k) = \sum_{k \geq 0} (n_{k+1} - n_k) P(T_{\tau_{\mathcal{N}}} > n_k)$$

with the convention $n_0 := 0$.

(iv) This follows immediately from the proofs of (ii) and (iii).

$\square$

Proof *(Theorem 7.18)* In view of Theorem 7.8 and Proposition 7.11 it suffices to show that there exists an invariant probability measure for (X_n).

First observe that we can always assume that $\xi(C) > 0$, where ξ is the minorizing measure of R_a. Indeed, let $\xi_k(\cdot) = a^k \int \xi(dy) P^k(y, \cdot)$. Then for all $x \in C$

$$R_a(x, \cdot) \geq a^k R_a P^k(x, \cdot) \geq \xi_k(\cdot)$$

so that ξ_k is another minorizing measure. Now, there exists k such that $\xi_k(C) > 0$, for otherwise we would have $P^k(y, C) = 0$ for all k and ξ-almost all y, in contradiction with the assumption that C is recurrent. Replacing ξ by such a ξ_k proves our claim.

Let $\tau_C < \tau_C^{(2)} < \tau_C^{(3)} < \ldots$ be the successive times at which (X_n) enters C, i.e., $\tau_C^{(k+1)} = \min\{n > \tau_C^{(k)} : X_n \in C\}$. By assumption (iii) (of the theorem to be proved) and the strong Markov property,

$$\mathbb{E}_x(\tau_C^{(k)}) \leq kM$$

for all $x \in C$. Let (Y_n) be the chain with kernel R_a, $\tau_C^Y = \min\{n \geq 1 : Y_n \in C\}$, and $Q(x, \cdot)$ the kernel on C defined by $Q(x, A) = \mathbb{P}_x(Y_{\tau_C^Y} \in A)$ for all Borel sets $A \subset C$. By Proposition 7.20 (i), $\tau_C^Y < \infty$ a.s. so that Q is a Markov kernel (i.e., $Q(x, C) = 1$). Furthermore $Q(x, A) \geq R_a(x, A) \geq \varepsilon \psi(A)$ with $\varepsilon = \xi(C)$ and $\psi(A) = \frac{\xi(A)}{\varepsilon}$. In other words, Q is a Markov kernel whose full state space (here C) is a small set. Then, by a theorem that will be proved later (Theorem 8.7 in Chap. 8), Q has a (unique) invariant probability measure π. If Y_0 is distributed according to π so is $Y_{\tau_C^Y}$, and by Proposition 7.20 (iii)

$$\mathbb{E}_\pi (\tau_C^Y) \leq \frac{M}{a}.$$

By Exercise 4.24 this implies that (Y_n) (or equivalently R_a), hence (X_n), admits an invariant probability measure.   □

7.6 Positive Recurrence for Feller Chains

The next results give some (much more tractable) conditions ensuring that a Feller chain is positive recurrent.

Theorem 7.21 *Let P be Feller. Assume that there exist a compact recurrent set C such that $\sup_{x \in C} \mathbb{E}_x(\tau_C) < \infty$ and an accessible weak Doeblin point $x^* \in \text{Int}(C)$ (the interior of C). Then the equivalent conditions of Theorem 7.8 hold true.*

Proof By assumption there exist a neighborhood $U \subset C$ of x^* and a nontrivial measure ξ such that $R_a(x, \cdot) \geq \xi(\cdot)$ for all $x \in U$. By Proposition 7.17, U is recurrent and $\sup_{x \in C} \mathbb{E}_x(\tau_U) < \infty$. We can then apply Theorem 7.18, with U in place of C. This proves the result.   □

Corollary 7.22 *Let P be Feller. Assume that there exist an accessible weak Doeblin point, a proper map $V : M \to \mathbb{R}_+$, and a nonnegative constant R such that $PV \leq V - 1$ on $\{V > R\}$ and $\sup_{\{x \in M : V(x) \leq R\}} PV(x) < \infty$. Then the equivalent conditions of Theorem 7.8 hold true.*

Proof Let x^* be the accessible weak Doeblin point. Choose R large enough so that $V(x^*) < R$. Set $C = \{V \leq R\}$ and apply Proposition 7.12 (a) and Theorem 7.21.   □

Theorem 7.23 *Let P be Feller. Assume that there exists an accessible weak Doeblin point and that for all $x \in M$ the family of empirical occupation measures (ν_n) is $\mathbb{P}_x$-almost surely tight (this is true for instance under the assumptions of Corollary 7.22). Then the equivalent conditions of Theorem 7.8 hold true.*

Proof By assumption there exists an open accessible petite set C. By Theorem 6.2 and Theorem 4.20, there exists a unique invariant probability measure π for P and $\nu_n \Rightarrow \pi$, $\mathbb{P}_x$-almost surely, for all $x \in M$. Since C is open and accessible, $\pi(C) > 0$ (see Proposition 5.8 (ii)) and, by the Portmanteau theorem, $\liminf \nu_n(C) \geq \pi(C)$. This proves that every point x leads almost surely to C. The result then follows from Proposition 7.11 and Theorem 7.8. $\qquad\qquad\square$

7.6.1 Application to PDMPs

Let $E = \{1, \dots, N\}$ be a set of environments and $\{G_i\}_{i \in E}$ a family of smooth globally integrable vector fields on $\mathbb{R}^k$.

Consider the PDMP $Z_t = (Y_t, I_t) \in \mathbb{R}^k \times E$ as defined in Sect. 6.4. Recall that, starting from $Z_0 = (Y_0, I_0) = (x, i) \in \mathbb{R}^k \times E$, Y_t follows the flow induced by the vector field G_i during a time τ_1 having an exponential distribution with parameter λ_i and $I_t = i$ on $[0, \tau_1)$. Then a new environment $j \in E$ is chosen with probability $p_j > 0$, Y_t follows the flow induced by G_j during a time $\tau_2 - \tau_1$ having an exponential distribution with parameter λ_j, and $I_t = j$ on $[\tau_1, \tau_2)$, etc.

If now the initial environment I_0 is randomly chosen with law $\sum_{i \in E} p_i \delta_i$, then $X_n = Y_{\tau_n}$ defines a Markov chain on $\mathbb{R}^k$, as explained in Sect. 6.3, whose kernel is given as (see formula (6.3))

$$Pf(x) = \sum_{i \in E} p_i \int_0^\infty f(\Phi_i(t, x)) \lambda_i e^{-\lambda_i t} \, dt. \qquad (7.5)$$

The following exercise shows that if there exists a common Lyapunov function for some (not necessarily all) of the vector fields G_i and if this function does not grow too fast along the flows of the other vector fields, then it can serve as a Lyapunov function for P.

Exercise 7.24 (Lyapunov Function for PDMPs)

(i) Suppose that there exists a proper C^1-map $V : \mathbb{R}^k \to \mathbb{R}_+$ and numbers $\alpha_1, \dots, \alpha_N$ (not necessarily negative) such that for each $i \in E$,

$$\limsup_{\|x\| \to \infty} \frac{\langle \nabla V(x), G_i(x) \rangle}{V(x)} \leq \alpha_i < \lambda_i.$$

Show that, if

$$\sum_{i \in E} p_i \frac{\lambda_i}{\lambda_i - \alpha_i} < 1,$$

then

$$PV \leq \rho V + \kappa$$

for some $0 \leq \rho < 1$ and $\kappa \geq 0$.

(ii) Let $\alpha_i(x)$ be the largest eigenvalue of the symmetric matrix $DG_i(x) + DG_i(x)^\top$ and let $\alpha_i = \sup_x \alpha_i(x)$. Show that

$$\limsup_{\|x\| \to \infty} \frac{\langle x, G_i(x) \rangle}{\|x\|^2} = \limsup_{\|x\| \to \infty} \frac{\langle x, G_i(x) - G_i(0) \rangle}{\|x\|^2} \leq \frac{\alpha_i}{2}.$$

Using (i), give conditions on $\alpha_1, \ldots, \alpha_N$ ensuring that $V(x) = \|x\|^2$ is a Lyapunov function for P.

(iii) Suppose that there exists a proper C^1-map $W : \mathbb{R}^k \to \mathbb{R}_+$ and numbers $a_1, \ldots, a_N$ (not necessarily negative) such that for each $i \in E$,

$$\limsup_{\|x\| \to \infty} \langle \nabla W(x), G_i(x) \rangle \leq a_i.$$

Show that, if

$$\sum_{i \in E} p_i \frac{a_i}{\lambda_i} < 0,$$

then, for $\varepsilon > 0$ sufficiently small, the map $V = e^{\varepsilon W}$ and the numbers $\alpha_i = \varepsilon a_i$ satisfy the conditions given in (i).

Theorem 7.25 *Suppose that there exists V as in Exercise 7.24 (i) and an accessible point (in the sense of Proposition 6.20) at which the weak bracket condition as defined in Sect. 6.3.1 holds. Then the chain (X_n) is positive recurrent. Furthermore, the process (Z_t) is also positive recurrent in the sense that, for all $f : M \times E \to \mathbb{R}$ measurable and bounded,*

$$\lim_{t \to \infty} \frac{1}{t} \int_0^t f(Z_s) \, ds = \hat{\mu}(f)$$

almost surely, where $\hat{\mu}$ stands for the invariant probability measure of (Z_t).

Proof Positive recurrence of (X_n) follows from Corollary 7.22 and Theorem 6.16. We now prove the second statement. Let $f \in B(\mathbb{R}^k \times E)$ and $\mathcal{A} = \{\lim_{t \to \infty} \frac{1}{t} \int_0^t f(Z_s) \, ds = \hat{\mu}(f)\}$. In order to show that $\mathbb{P}_{x,i}(\mathcal{A}) = 1$ for all $(x, i) \in \mathbb{R}^k \times E$ it suffices to show that $\sum_{i \in E} p_i \mathbb{P}_{x,i}(\mathcal{A}) = 1$ for all x. This means one needs to show that $\lim_{t \to \infty} \frac{1}{t} \int_0^t f(Z_s^x) \, ds \to \hat{\mu}(f)$ almost surely, where (Z_t^x) stands for the PDMP with initial condition (x, I_0) and I_0 has distribution $\sum_{i \in E} p_i \delta_i$.

Let $\mathcal{G}_n = \sigma\{(I_0, \tau_1), (I_{\tau_1}, \tau_2 - \tau_1), \ldots, (I_{\tau_{n-1}}, \tau_n - \tau_{n-1})\}$. Then $\int_0^{\tau_n} f(Z_s^x)\, ds$ is $\mathcal{G}_n$-measurable, and

$$\mathsf{E}\left(\int_{\tau_n}^{\tau_{n+1}} f(Z_s^x)\, ds \,\Big|\, \mathcal{G}_n \right) = \hat{f}(X_n^x),$$

where

$$\hat{f}(x) = \sum_{i \in E} p_i \int_0^\infty \int_0^t f(\Phi_i(s, x), i) \lambda_i e^{-\lambda_i s}\, ds\, dt$$

$$= \sum_{i \in E} p_i \int_0^\infty f(\Phi_i(t, x), i) e^{-\lambda_i t}\, dt.$$

Also,

$$\mathsf{Var}\left(\int_{\tau_n}^{\tau_{n+1}} f(Z_s^x)\, ds \,\Big|\, \mathcal{G}_n \right) \le \mathsf{E}((\tau_{n+1} - \tau_n)^2 | \mathcal{G}_n) \|f\|^2 \le \max_i \frac{2}{\lambda_i^2} \|f\|^2.$$

Thus, by the strong law of large numbers for martingales (see Theorem A.8),

$$\lim_{n \to \infty} \frac{1}{n} \left(\int_0^{\tau_n} f(Z_s^x)\, ds - \sum_{k=0}^{n-1} \hat{f}(X_k^x) \right) = 0$$

almost surely. On the other hand, by the strong law of large numbers, $\lim_{n \to \infty} \frac{\tau_n}{n} = \sum_{j \in E} p_j / \lambda_j$. Thus

$$\lim_{t \to \infty} \frac{1}{t} \int_0^t f(Z_s^x)\, ds = \frac{1}{\sum_{j \in E} p_j / \lambda_j} \mu(\hat{f}),$$

where μ is the invariant probability measure of (X_n). This proves the positive recurrence of (Z_t) and also gives - in this special case - an alternative proof of Theorem 6.26 (i). $\square$

7.6.2 Application to SDEs

Using the notation and assumptions of Sect. 6.5, consider the stochastic differential equation (6.12). Recall from the proof of Theorem 6.34 that the "formal" generator of (6.12) is the operator L defined on C^2-functions $f : \mathbb{R}^k \to \mathbb{R}$ by

$$Lf(x) = G_0(f)(x) + \frac{1}{2}\sum_{i=1}^{N} G_i^2(f)(x).$$

Lemma 7.26 *Suppose there exists a proper C^2-function $U : \mathbb{R}^k \to \mathbb{R}_+$ and positive numbers α, β such that*

$$LU \le -\alpha U + \beta.$$

Then, for all $t \ge 0$,

$$P_t U \le e^{-\alpha t} U + \frac{\beta}{\alpha}(1 - e^{-\alpha t}).$$

Proof Set $W_t = e^{\alpha t}(U(X_t^x) - \frac{\beta}{\alpha})$. By Itô's formula,

$$W_t - W_0 = \int_0^t e^{\alpha s}[\alpha U(X_s^x) - \beta + LU(X_s)]\, ds + M_t \le M_t,$$

where $(M_t)_{t\ge 0}$ is a local martingale with $M_0 = 0$. Thus, for all $n \in \mathbb{N}$, $(M_t^n) = (M_{t\wedge n})_{t\ge 0}$ is a continuous local martingale which is bounded below (by $-\frac{\beta}{\alpha}e^{\alpha n} - (U(x) - \frac{\beta}{\alpha})$). A local martingale that is bounded below may not be a martingale but is always a supermartingale (see, e.g., [45, Proposition 4.7]). Therefore $\mathsf{E}(W_{t\wedge n}) - U(x) + \frac{\beta}{\alpha} \le \mathsf{E}(M_t^n) \le \mathsf{E}(M_0) = 0$. Hence, $\mathsf{E}(U(X_{t\wedge n})) \le e^{-\alpha(t\wedge n)}(U(x) - \frac{\beta}{\alpha}) + \frac{\beta}{\alpha}$ and the desired result follows by Fatou's lemma. $\square$

Corollary 7.27 *Suppose there exists a proper C^2-function as in Lemma 7.26 for (6.12) and an accessible point p at which the weak Hörmander condition is satisfied. Then $\{P_t\}_{t\ge 0}$ has a unique invariant probability measure μ and for every $f \in B(\mathbb{R}^k)$ and $x \in \mathbb{R}^k$*

$$\lim_{t\to\infty} \frac{1}{t}\int_0^t f(X_s^x)\, ds = \mu(f).$$

Proof Let G be the 1-resolvent. Then, by Lemma 7.26, $GU \le \frac{1}{1+\alpha}U + \frac{\beta}{\alpha}$. By Corollary 7.22 and Theorem 6.34, G is a positive recurrent Markov kernel. The final statement follows from Proposition 4.58 (ii). $\square$

Chapter 8
Harris Ergodic Theorem

This chapter generalizes the convergence theorems proved in Chap. 2 for countable chains to general chains under the assumption that there exist an aperiodic small set (or Doeblin point) and, when M is noncompact, a Lyapunov function or a suitable control on the moments of the return time to this set. The final section discusses convergence in Wasserstein distance.

8.1 Total Variation Distance

Recall that $B(M)$ is the set of real-valued bounded measurable maps on M. Given two probability measures α and β on M, the *total variation distance* between α and β is defined by

$$|\alpha - \beta| = \sup\{|\alpha(f) - \beta(f)| \; : \; f \in B(M), \; \|f\|_\infty \leq 1\}. \tag{8.1}$$

See also Remark 5.21 in Sect. 5.3. It is easy to verify that the total variation distance defines a metric on $\mathcal{P}(M)$.

Note that if K is a Markov kernel on M,

$$|\alpha K - \beta K| \leq |\alpha - \beta| \tag{8.2}$$

because K maps $\{f \in B(M) : \|f\|_\infty \leq 1\}$ into itself.

Proposition 8.1 *Let* $\alpha, \beta \in \mathcal{P}(M)$.

(i)

$$|\alpha - \beta| = 2 \sup_{A \in \mathcal{B}(M)} \alpha(A) - \beta(A).$$

© The Author(s), under exclusive license to Springer Nature Switzerland AG 2022
M. Benaïm, T. Hurth, *Markov Chains on Metric Spaces*, Universitext,
https://doi.org/10.1007/978-3-031-11822-7_8

(ii) *Assume α and β are absolutely continuous with respect to $\gamma \in \mathcal{P}(M)$ with respective densities p and q. Then*

$$|\alpha - \beta| = \int |p - q| d\gamma.$$

(iii) *The space $\mathcal{P}(M)$ equipped with the total variation distance is complete.*

Proof We begin by proving assertion (ii). For all $f \in B(M)$ with $\|f\|_\infty \leq 1$, $|\alpha(f) - \beta(f)| \leq \int |p - q| d\gamma$ so that $|\alpha - \beta| \leq \int |p - q| d\gamma$. Conversely, set $f = \mathbf{1}_{p>q} - \mathbf{1}_{p<q}$. Then $\alpha(f) - \beta(f) = \int |p - q| d\gamma$.

We now pass to the proof of (i). We can always assume that for some $\gamma \in \mathcal{P}(M)$, α and β are absolutely continuous with respect to γ. It suffices for instance to choose $\gamma = \frac{\alpha+\beta}{2}$. Then

$$|\alpha - \beta| = \int_G (p - q) d\gamma + \int_{M \setminus G} (q - p) d\gamma = 2(\alpha(G) - \beta(G))$$

with $G = \{p > q\}$. Also, for all $A \in \mathcal{B}(M), \alpha(A) - \beta(A) \leq \alpha(A \cap G) - \beta(A \cap G) \leq \alpha(G) - \beta(G)$. Our last task is to prove completeness. Let (μ_n) be a Cauchy sequence for the total variation distance. Then, in view of (i), for every Borel set A, $(\mu_n(A))$ is a Cauchy sequence in $\mathbb{R}$, hence converges to some number $\mu(A)$. By the Cauchy property, the convergence is uniform in A, i.e., $\sup_{A \in \mathcal{B}(M)} |\mu_n(A) - \mu(A)| \to 0$. From this it is easy to verify that μ is a probability measure on M. $\square$

Exercise 8.2 For $f : M \to \mathbb{R}$, let $\Delta(f) = \sup\{\frac{|f(x)-f(y)|}{2} : x, y \in M\}$. Show that

$$|\alpha - \beta| = \sup\{|\alpha(f) - \beta(f)| : f \text{ measurable}, \Delta(f) \leq 1\}.$$

Remark 8.3 Although the total variation distance (8.1) and the Fortet-Mourier distance (4.2) look very similar, they induce quite different topologies on $\mathcal{P}(M)$. Clearly,

$$\rho(\alpha, \beta) \leq |\alpha - \beta|$$

so that convergence in total variation implies weak convergence; but the converse is false. Let, for example, X be a random variable on $\mathbb{R}$ whose law P_X is absolutely continuous with respect to the Lebesgue measure dx (e.g., a Gaussian random variable) and $X_n = \frac{X}{n}$. Then $X_n \to 0$ almost surely, hence $\mathsf{P}_{X_n} \Rightarrow \delta_0$, while $|\mathsf{P}_{X_n} - \delta_0| = 2$ by Proposition 8.1 (i).

Remark 8.4 (Total Variation of Signed Measures) A *finite signed measure* on M is a map $\mu : \mathcal{B}(M) \to \mathbb{R}$ such that $\mu(\emptyset) = 0$ and which is σ-additive, meaning that

$$\mu\left(\bigcup_n A_n\right) = \sum_n \mu(A_n)$$

for any family $\{A_n\}$, $A_n \in \mathcal{B}(M)$, having disjoint elements. The Hahn-Jordan decomposition theorem (see [21], Theorem 5.6.1) asserts that such a measure can be written as

$$\mu = \mu^+ - \mu^-,$$

where μ^+ and μ^- are nonnegative measures that are mutually singular: There exists $D \in \mathcal{B}(M)$ such that for all $A \in \mathcal{B}(M)$, $\mu^+(A) = \mu(A \cap D)$ and $\mu^-(A) = -\mu(A \cap D^c)$. The *total variation measure* of μ is the nonnegative measure $\mu^+ + \mu^-$ and its *total variation norm* is

$$|\mu| = \mu^+(M) + \mu^-(M) = \sup\{|\mu(f)| \ : \ f \in B(M), \ \|f\|_\infty \leq 1\}.$$

When M is a compact metric space, the topological dual $C^*(M)$ of $C(M)$ can be identified with the space of bounded signed measures equipped with the total variation norm, so that convergence in total variation coincides with (strong) convergence in $C^*(M)$. We refer the reader to [21], Chapter 7, for more details and a proof of this latter point.

Exercise 8.5 Use the Hahn-Jordan decomposition to show assertion (*i*) of Proposition 8.1.

8.1.1 Coupling

Given $\alpha, \beta \in \mathcal{P}(M)$, a *coupling* of α and β is a random vector (X, Y) defined on some probability space $(\Omega, \mathcal{F}, \mathsf{P})$ taking values in $M \times M$ such that X has distribution α and Y has distribution β.

Proposition 8.6 *Let* $\alpha, \beta \in \mathcal{P}(M)$. *Then*

(i) *(Coupling inequality) For every coupling* (X, Y) *of* (α, β),

$$|\alpha - \beta| \leq 2\mathsf{P}(X \neq Y);$$

(ii) *(Maximal coupling) There exists a coupling* (X, Y) *of* (α, β) *such that*

$$|\alpha - \beta| = 2\mathsf{P}(X \neq Y).$$

Proof

(i) For all $A \in \mathcal{B}(M)$,

$$P(X \in A) - P(Y \in A) = P(X \in A; X \neq Y) - P(Y \in A; X \neq Y) \leq P(X \neq Y).$$

This inequality, combined with Proposition 8.1 (i), proves (i).

(ii) Assume without loss of generality that $d\alpha = p\,d\gamma$ and $d\beta = q\,d\gamma$ for some $\gamma \in \mathcal{P}(M)$ (e.g., $\gamma = (\alpha + \beta)/2$). Then, by Proposition 8.1 (ii),

$$|\alpha - \beta| = \int |p - q|\,d\gamma = 2(1 - \varepsilon),$$

where $\varepsilon = \int (p \wedge q)\,d\gamma$. If $\varepsilon = 0$, α and β are mutually singular and any coupling satisfies the equality $|\alpha - \beta| = 2P(X \neq Y) = 2$. If $\varepsilon \neq 0$, let $U \in M, V \in M, W \in M, \Theta \in \{0, 1\}$ be independent random variables having distributions $\frac{1}{\varepsilon}(p \wedge q)\,d\gamma, \frac{1}{1-\varepsilon}(p - (p \wedge q))\,d\gamma, \frac{1}{1-\varepsilon}(q - (p \wedge q))\,d\gamma$, and $(1-\varepsilon)\delta_0+\varepsilon\delta_1$, respectively. Set $X = \Theta U+(1-\Theta)V$ and $Y = \Theta U+(1-\Theta)W$. Then $P(X \neq Y) = P(\Theta = 0) = (1 - \varepsilon)$ and (X, Y) is a coupling of (α, β). $\square$

8.2 Harris Convergence Theorems

Throughout this section P is a Markov kernel on M. Recall from Chap. 6 that a set $C \in \mathcal{B}(M)$ is called a *small set* for P if there exists a nontrivial measure ξ on M (called the minorizing measure of C) such that

$$P(x, \cdot) \geq \xi(\cdot) \tag{8.3}$$

for all $x \in C$. Recall also that a point in M is called a Doeblin point if it has a neighborhood which is a small set.

8.2.1 Geometric Convergence

The importance of small sets is emphasized by the following simple version of Harris's theorem (sometimes called Doeblin's theorem).

Theorem 8.7 *Let $m \in \mathbb{N}^*$. Suppose that M is a small set for P^m with minorizing measure ξ. Then, for all $\alpha, \beta \in \mathcal{P}(M)$,*

$$|\alpha P^n - \beta P^n| \leq (1 - \varepsilon)^{\lfloor n/m \rfloor}|\alpha - \beta|,$$

where $0 < \varepsilon = \xi(M) \le 1$. *Furthermore* P *has a unique invariant probability measure* π *and*

$$|\alpha P^n - \pi| \le (1 - \varepsilon)^{[n/m]}|\alpha - \pi|.$$

Proof First suppose $m = 1$. Set $\psi = \frac{\xi}{\xi(M)}$, $\varepsilon = \xi(M)$, and

$$K(x, \cdot) = \frac{P(x, \cdot) - \varepsilon \psi(\cdot)}{1 - \varepsilon} \quad \text{if } \varepsilon < 1.$$

Then K is a Markov kernel and $\alpha P = \varepsilon \psi + (1 - \varepsilon)\alpha K$ so that

$$|\alpha P - \beta P| = (1 - \varepsilon)|\alpha K - \beta K| \le (1 - \varepsilon)|\alpha - \beta|,$$

where the last inequality follows from (8.2). Hence, $\alpha \mapsto \alpha P$ is a strict contraction for the total variation distance. Then

$$|\alpha P^n - \beta P^n| \le (1 - \varepsilon)^n |\alpha - \beta|$$

and $\alpha \mapsto \alpha P$ has a unique fixed point, by application of the Banach fixed point theorem, because the space of probability measures endowed with the total variation distance is complete.

If now $m \ge 1$, set $Q = P^m$. Write $n = km + r$ for $r \in \{0, \ldots, m - 1\}$ and

$$|\alpha P^n - \beta P^n| = |\alpha P^r Q^k - \beta P^r Q^k| \le (1 - \varepsilon)^k |\alpha P^r - \beta P^r| \le (1 - \varepsilon)^k |\alpha - \beta|.$$

To conclude, recall that if π is invariant for P^m, then $\frac{1}{m} \sum_{k=0}^{m-1} \pi P^k$ is invariant for P.   □

Remark 8.8 Theorem 8.7 is purely measure-theoretic and does not require that M is a metric space.

Aperiodic Small Sets

A measurable set $C \subset M$ is said to be *aperiodic* if the set

$$R(C) = \{k \ge 1 : \inf_{x \in C} P^k(x, C) > 0\}$$

is nonempty and aperiodic as defined in Sect. 2.4.

Exercise 8.9

(a) Let P be Feller and let $U \subset M$ be an open, accessible (i.e., $R_a(x, U) > 0$ for all $x \in M$) small set. Show that $R(U)$ is nonempty.

(b) Construct a Feller Markov chain having an open recurrent set U for which $R(U) = \emptyset$. *Hint:* Let $\{\Phi_t\}$ be the flow on $S^1 = \mathbb{R}/2\pi\mathbb{Z}$ induced by the differential equation $\dot{\theta} = \sin^2(\theta/2)$. Consider the deterministic chain defined as $X_n^x = \Phi_n(x)$. One can show that every proper neighborhood U of 0 is recurrent, but $R(U) = \emptyset$.

Let $x^* \in M$ be an accessible Doeblin point for P Feller. We say that x^* is *aperiodic* if it has a neighboring small set U which is aperiodic. Observe that if U is a neighboring small set of x^* such that $\xi(U) > 0$ (where ξ stands for the minorizing measure of U) then x^* is aperiodic.

Proposition 8.10 *Assume P is Feller. Let $x^* \in M$ be an accessible and aperiodic Doeblin point and let $C \subset M$ be a compact set. Then there exists $m \geq 1$ such that C is a small set for P^m.*

Proof Let U be an open neighboring small set of x^* with $R(U)$ aperiodic. Then, by aperiodicity, there exists $n_0 \in \mathbb{N}$ such that $k \in R(U)$ for all $k \geq n_0$ (see Proposition 2.21).

For $\delta > 0$ and $k \in \mathbb{N}^*$ let $O(\delta, k) = \{x \in M : P^k(x, U) > \delta\}$. By Feller continuity and the Portmanteau theorem 4.1, $O(\delta, k)$ is an open set. Since x^* is accessible, the family $\{O(\delta, k) : \delta > 0, k \in \mathbb{N}^*\}$ covers M. Thus, by compactness, there exist $\delta > 0$ and integers $k_1, \ldots, k_n$ such that $C \subset \cup_{i=1}^n O(\delta, k_i)$. For $x \in O(\delta, k_i)$ and $k > n_0$,

$$P^{k_i+k}(x, .) \geq \int_U P^{k_i}(x, dy) P^k(y, .)$$

$$\geq \int_U P^{k_i}(x, dy) P^{k-1}(y, U)\xi(.) \geq \delta \inf_{y \in U} P^{k-1}(y, U)\xi(.).$$

Here ξ stands for the minorizing measure of U. Thus, for $m = \max\{k_1, \ldots, k_n\} + n_0 + 1$ and some $\delta' > 0$,

$$\inf_{x \in C} P^m(x, .) \geq \delta'\xi(.).$$

$\square$

Theorem 8.7 and Proposition 8.10 imply the following useful result for Feller chains on compact sets.

Corollary 8.11 *Assume P is Feller on M compact and that there exists an accessible and aperiodic Doeblin point. Then the conclusion of Theorem 8.7 holds.*

When M is not compact, the assumption (made in Theorem 8.7 or used in Corollary 8.11) that the whole space is a small set is usually not satisfied. A sufficient condition ensuring geometric convergence is the existence of a small set and a Lyapunov function forcing the system to enter this small set. A classical proof relying on coupling and renewal properties will be given in the next section. Hairer and Mattingly in [36] gave an alternative beautiful proof based on the construction of a suitable semi-norm making P a strict contraction. This proof is given below.

Theorem 8.12 (Harris, Hairer and Mattingly) *Assume that there exist:*

(a) *A measurable map* $V : M \to \mathbb{R}_+, 0 < \rho < 1,$ *and* $\kappa \geq 0$ *such that*

$$PV \leq \rho V + \kappa;$$

(b) *A probability measure* ψ *on* M *and* $0 < \varepsilon \leq 1$ *such that*

$$P(x, \cdot) \geq \varepsilon \psi(\cdot)$$

for all $x \in V_R := \{x \in M : V(x) \leq R\}$ *and* $R \geq 2\kappa/(1 - \rho).$

Then there exist a unique invariant probability measure π *for* P *and constants* $0 \leq \gamma < 1, C > 0$ *such that for all* $f : M \to \mathbb{R}$ *measurable with* $\|f\|_V := \sup_{x \in M} \frac{|f(x)|}{1+V(x)} < \infty,$

$$|P^n f(x) - \pi(f)| \leq C\gamma^n (1 + V(x))\|f\|_V$$

for all $x \in M$ *and* $n \in \mathbb{N}^*.$

Proof For $\beta > 0$ and $f : M \to \mathbb{R}$ measurable, possibly unbounded, let

$$\|f\|_\beta = \sup\left\{ \frac{|f(x) - f(y)|}{2 + \beta(V(x) + V(y))} : x, y \in M \right\}.$$

We claim that for some $1 \geq \beta > 0$ and $0 \leq \gamma < 1,$

$$\|f\|_\beta \leq 1 \Rightarrow \|Pf\|_\beta \leq \gamma. \tag{8.4}$$

Assume the claim is proved. Observe that $\|f\|_1 \leq \|f\|_\beta \leq \frac{1}{\beta}\|f\|_1 \leq \frac{1}{\beta}\|f\|_V.$ Then

$$\|P^n f\|_1 \leq \|P^n f\|_\beta \leq \gamma^n \|f\|_\beta \leq \gamma^n \beta^{-1}\|f\|_V.$$

Equivalently

$$|P^n f(x) - P^n f(y)| \leq \gamma^n \beta^{-1}\|f\|_V (2 + V(x) + V(y)), \quad x, y \in M.$$

Thus,

$$|P^n f(x) - \pi f| \leq \int |P^n f(x) - P^n f(y)|\pi(dy) \leq \gamma^n \beta^{-1}\|f\|_V (2 + V(x) + \pi V),$$

where π is some (hence unique) invariant probability measure (see Exercise 8.13). This proves the result.

We now prove the claim. Let f be such that $\|f\|_\beta \le 1$ and let $x, y \in M$. Suppose first that $V(x) + V(y) > R$. Then

$$|Pf(x) - Pf(y)| = |\int (f(u) - f(v))\delta_x P(du)\delta_y P(dv)|$$

$$\le \int |f(u) - f(v)|\delta_x P(du)\delta_y P(dv) \le 2 + \beta PV(x) + \beta PV(y)$$

$$\le 2 + 2\beta\kappa + \rho\beta(V(x) + V(y)) \le \gamma_1(2 + \beta(V(x) + V(y))),$$

where

$$\gamma_1 = \frac{\beta(2\kappa + \rho R) + 2}{\beta R + 2} \in (0, 1).$$

The last inequality follows from the fact that for all $\rho, r > 0$ and $a \ge 2\rho$,

$$t \ge r \Rightarrow a + \rho t \le \gamma_1(2 + t),$$

where γ_1 is the solution to $a + \rho r = \gamma_1(2 + r)$. It suffices to set $a = 2 + 2\beta\kappa$ and $r = \beta R$.

Suppose now that $V(x) + V(y) \le R$. In particular, $x, y \in V_R$. As in the proof of Theorem 8.7, write $Pf = (1 - \varepsilon)Kf + \varepsilon\psi(f)$, where, for all $x \in V_R$, $K(x, \cdot)$ is a Markov operator. Thus

$$|Pf(x) - Pf(y)| = (1 - \varepsilon)|Kf(x) - Kf(y)| \le (1 - \varepsilon)(2 + \beta(KV(x) + KV(y))).$$

Also, $(1 - \varepsilon)KV(x) = PV(x) - \varepsilon\psi V \le \rho V(x) + \kappa$. Thus

$$|Pf(x) - Pf(y)| \le 2(1 - \varepsilon) + 2\beta\kappa + \rho\beta(V(x) + V(y)) \le \gamma_2(2 + \beta(V(x) + V(y)))$$

with $\gamma_2 = \max(\rho, 1 - \varepsilon + \beta\kappa)$. Finally it suffices to choose $\beta\kappa < \varepsilon$ and to set $\gamma = \max(\gamma_1, \gamma_2)$.   □

Exercise 8.13

(i) Suppose that M is a Polish space, P is Feller, and that there exists a proper and continuous map $V : M \to \mathbb{R}_+$ satisfying assumption (a) of Theorem 8.12. Show that the set $\mathsf{Inv}(P)$ is nonempty. *Hint:* Use Corollary 4.23.

(ii) Suppose only that M is a measurable space. Show that $\mathcal{P}_V(M) = \{\mu \in \mathcal{P}(M) : V \in L^1(\mu)\}$ is complete for the distance

$$|\mu - v|_\beta := \sup\{|\mu f - vf| : f : M \to \mathbb{R} \text{ measurable}, \|f\|_\beta \le 1\}.$$

Deduce that, under the assumptions of Theorem 8.12, there exists a unique invariant probability measure for P. *Hint:* Use Inequality (8.4) to show that

$$|\mu P - vP|_\beta \le \gamma|\mu - v|_\beta \qquad (8.5)$$

for some $0 \le \gamma < 1$ and $\beta > 0$.

Corollary 8.14 *Suppose P is Feller and that there exists a proper map $V : M \to \mathbb{R}_+$ satisfying assumption (a) of Theorem 8.12. Suppose furthermore that there exists an accessible aperiodic Doeblin point. Then the conclusion of Theorem 8.12 holds true.*

Proof Choose $R > \frac{2\kappa}{(1-\rho)^2}$. The set $C = \{V \le R\}$ is a compact set (because V is proper) and small for some P^m by Proposition 8.10. Since $P^m V \le \rho^m V + \frac{\kappa}{1-\rho}$, Theorem 8.12 applies to P^m and the result follows.   □

8.2.2 Continuous Time: Exponential Convergence

For a weak Feller continuous-time Markov process $\{P_t\}_{t \ge 0}$, aperiodicity is not an issue. Indeed, if a point $p \in M$ is accessible for $\{P_t\}_{t \ge 0}$ and is a Doeblin point for some P_{T_0}, then p is necessarily aperiodic for P_{T_0}. This is a direct consequence of Lemma 6.5. Thus, the continuous-time version of Corollary 8.14 reads as follows:

Theorem 8.15 *Let $\{P_t\}_{t \ge 0}$ be a continuous-time weak Feller semigroup. Assume in addition the following:*

(i) *There exists a point $p \in M$ which is accessible for $\{P_t\}_{t \ge 0}$ and which is a Doeblin point for some P_{T_0} with $T_0 > 0$;*
(ii) *There exist a proper map $V : M \to \mathbb{R}_+$, $0 \le \rho < 1$, $\kappa \ge 0$, and $T_1 > 0$ such that*

$$P_{T_1} V \le \rho V + \kappa.$$

Then there exist a unique invariant probability measure π for $\{P_t\}_{t \ge 0}$ and constants $a > 0, C > 0$ such that for all $f : M \to \mathbb{R}$ measurable,

$$|P_t f(x) - \pi(f)| \le Ce^{-at}(1 + V(x))\|f\|_V$$

for all $x \in M$ and $t \ge 0$.

Proof Relying on Proposition 6.3, one can find a point q and a time $T = mT_1$ (with $m \in \mathbb{N}^*$ sufficiently large) such that q is an accessible Doeblin point for P_T. By Lemma 6.5, it is also aperiodic for P_T. By assumption (ii), $P_T V \leq \rho^k + \frac{\kappa}{1-\rho}$. Thus, by Corollary 8.14, there exist constants $0 < \gamma < 1$ and $C \geq 0$ such that for all $n \in \mathbb{N}$ and $0 \leq r < T$,

$$|P_{nT+r} f(x) - \pi(f)| = |P_T^n P_r f(x) - \pi(P_r f)| \leq C\gamma^n (1 + V(x))\|f\|_V.$$

Thus

$$|P_t f(x) - \pi(f)| \leq \frac{C}{\gamma} e^{\log(\gamma)T/t} (1 + V(x))\|f\|_V$$

for all $x \in M$ and $t \geq 0$.   □

Example 8.16 (Piecewise Deterministic Markov Processes) Consider the piecewise deterministic Markov process defined in Sect. 6.4. Suppose that there exist an accessible point at which the strong bracket condition holds and a Lyapunov function as in Exercise 7.24. Then the conclusions of Theorem 8.15 hold.

Example 8.17 (Stochastic Differential Equations) Consider the stochastic differential equation introduced in Sect. 6.5. Suppose that there exist an accessible point at which the Hörmander condition holds and a Lyapunov function as in Lemma 7.26. Then the conclusions of Theorem 8.15 hold.

8.2.3   Coupling, Splitting, and Polynomial Convergence

This section is the natural counterpart of Sect. 2.7 on countable chains. It revisits the convergence theorems of the previous section and relates the rate of convergence to the moments of the return time to a recurrent small set.

Theorem 8.18 *Let $C \subset M$ be an aperiodic, recurrent small set for P.*

(i) *If $\sup_{x \in C} \mathbb{E}_x(\tau_C) < \infty$, then P is positive recurrent and, letting π denote its invariant probability measure,*

$$\lim_{n \to \infty} |\mu P^n - \pi| = 0$$

for every $\mu \in \mathcal{P}(M)$.

(ii) *If* $\sup_{x \in C} \mathbb{E}_x(\tau_C^p) < \infty$ *for some* $p \geq 2$, *then there exists* $c \geq 0$ *such that for every* $\mu \in \mathcal{P}(M)$ *and for every* $n \in \mathbb{N}^*$,

$$|\mu P^n - \pi| \leq \frac{1}{n^{p-1}} c(1 + \mathbb{E}_\mu(\tau_C^{p-1})).$$

(iii) *If* $\sup_{x \in C} \mathbb{E}_x(e^{\lambda_0 \tau_C}) < \infty$ *for some* $\lambda_0 > 0$, *then there exist* $0 < \lambda < \lambda_0$ *and* $c \geq 0$ *such that for every* $\mu \in \mathcal{P}(M)$ *and for every* $n \in \mathbb{N}^*$,

$$|\mu P^n - \pi| \leq e^{-\lambda n} c(1 + \mathbb{E}_\mu(e^{\lambda_0 \tau_C})).$$

Proof Positive recurrence follows from Theorem 7.18. The rest of the proof relies on a coupling argument that goes back to Harris [37] and Nummelin [52]. Let C be an aperiodic recurrent set for P. We proceed in two steps.

Step 1 We first assume that C is an *atom*, meaning that there exists a probability measure ξ on M such that for all $x \in C$, $P(x, \cdot) = \xi(\cdot)$. In this situation the proof is very much like the proof given for a countable Markov chain (Theorem 2.35). Let (X_n) and (Y_n) be two independent chains (induced by P), $\mathbb{P}_{\mu \otimes \nu}$ the law of $((X_n, Y_n))_{n \geq 0}$ when (X_0, Y_0) has law $\mu \otimes \nu$, and let

$$\tau_{C \times C} = \min\{n \geq 1 : X_n \in C, Y_n \in C\}.$$

Since C is an atom, for all $\mu, \nu \in \mathcal{P}(M)$ and $n \in \mathbb{N}^*$,

$$\mathbb{P}_{\mu \otimes \nu}(X_n \in \cdot; \tau_{C \times C} < n) = \mathbb{P}_{\mu \otimes \nu}(Y_n \in \cdot; \tau_{C \times C} < n).$$

Hence

$$|\mu P^n - \pi| = |\mu P^n - \pi P^n| \leq \mathbb{P}_{\mu \otimes \pi}(\tau_{C \times C} \geq n), \tag{8.6}$$

where π is the unique invariant probability measure of P. Let now $(\tau_C^{(n)})$ (respectively $(\tilde{\tau}_C^{(n)})$) denote the successive hitting times of C by (X_n) (respectively (Y_n)). The assumption that C is an aperiodic atom makes the processes $T := (\tau_C^{(n+1)})_{n \geq 0}$ and $\tilde{T} := (\tilde{\tau}_C^{(n+1)})_{n \geq 0}$ two aperiodic independent renewal processes (see Sect. 2.6) and $\tau_{C \times C}$ is their first common renewal time. The additional assumption that $\sup_{x \in C} \mathbb{E}_x(\tau_C) < \infty$ makes these processes L^1 (as defined in Sect. 2.6) so that $\tau_{C \times C} < \infty$ almost surely (see Equation (2.5) and the discussion preceding it). Together with (8.6), this proves the first assertion. To prove the second assertion, observe that by (8.6), Markov's inequality, and Theorem 2.33, one has for all $0 < q \leq p$ that

$$|\mu P^n - \pi| \leq \frac{1}{n^q} \mathbb{E}_{\mu \otimes \pi}(\tau_{C \times C}^q) \leq \frac{1}{n^q} c(1 + \mathbb{E}_\mu(\tau_C^q) + \mathbb{E}_\pi(\tau_C^q)).$$

The problem then reduces to estimating $\mathbb{E}_\pi(\tau_C^q)$. Here again, the assumption that C is an atom will prove to be very useful. As for countable Markov chains, π can be explicitly written as

$$\pi(f) = \frac{\mathbb{E}_x(f(X_1) + \ldots + f(X_{\tau_C}))}{\mathbb{E}_x(\tau_C)} = \pi(C)\mathbb{E}_x(f(X_1) + \ldots + f(X_{\tau_C}))$$

for any $x \in C$ and all $f \geq 0$ measurable. The proof is similar to the proof of assertion (iii) in Theorem 2.6 (compare to Exercise 4.24) and left to the reader. Applying this formula to the map $y \mapsto \mathbb{E}_y(\psi(\tau_C))$ for some nonnegative function ψ leads to

$$\mathbb{E}_\pi(\psi(\tau_C)) = \pi(C)\mathbb{E}_x\left(\sum_{k=0}^{\tau_C-1}\psi(k)\right),$$

for all $x \in C$, exactly as in Proposition 2.10. In particular

$$\mathbb{E}_\pi(\tau_C^q) \leq \pi(C)\mathbb{E}_x(\tau_C^{q+1})$$

for all $x \in C$. With $q = p - 1$, this estimate yields

$$\mathbb{E}_\pi(\tau_C^{p-1}) \leq \pi(C)\sup_{x \in C}\mathbb{E}_x(\tau_C^p) < \infty,$$

which concludes the proof of the second assertion.

The proof of the third assertion is similar. By Markov's inequality and Theorem 2.34 there exists $0 < \lambda \leq \lambda_0$ such that

$$|\mu P^n - \pi| \leq e^{-\lambda n}\mathbb{E}_{\mu \otimes \pi}(e^{\lambda \tau_{C \times C}}) \leq e^{-\lambda n}c(1 + \mathbb{E}_\mu(e^{\lambda_0 \tau_C}) + \mathbb{E}_\pi(e^{\lambda_0 \tau_C})).$$

And for all $x \in C$,

$$\mathbb{E}_\pi(e^{\lambda_0 \tau_C}) = \pi(C)\mathbb{E}_x\left(\frac{e^{\lambda_0 \tau_C} - 1}{e^{\lambda_0} - 1}\right).$$

Step 2 We suppose now that C is a small set with minorizing measure ξ. Let $\varepsilon = \xi(M) < 1$, $\psi(\cdot) = \frac{\xi(\cdot)}{\varepsilon}$, and let K be the kernel on C defined by

$$K(x, \cdot) = \frac{P(x, \cdot) - \varepsilon\psi(\cdot)}{1 - \varepsilon}.$$

The idea of the splitting method consists in constructing a Markov chain (X_n) with kernel P with the help of an auxiliary sequence (I_n), $I_n \in \{0, 1\}$. If $X_n \notin C$, then I_n

is set to 0. If $X_n \in C$, I_n is randomly chosen according to a Bernoulli distribution with parameter ε. At the next step, X_{n+1} is distributed according to

$$P(X_n, \cdot)\mathbf{1}_{\{X_n \in M \setminus C\}} + [(1 - I_n)K(X_n, \cdot) + I_n \psi(\cdot)]\mathbf{1}_{\{X_n \in C\}}.$$

More formally, consider the Markov kernel Q defined on

$$\mathcal{M} = \{(x, i) \in M \times \{0, 1\} : x \notin C \Rightarrow i = 0\}$$

as follows: For all $x \in M \setminus C$,

$$Q(x, 0; dy \times \{0\}) = P(x, dy)(1 - \varepsilon \mathbf{1}_C(y)),$$
$$Q(x, 0; dy \times \{1\}) = P(x, dy)\varepsilon \mathbf{1}_C(y),$$

and for all $x \in C$,

$$Q(x, 0; dy \times \{0\}) = K(x, dy)(1 - \varepsilon \mathbf{1}_C(y)),$$
$$Q(x, 0; dy \times \{1\}) = K(x, dy)\varepsilon \mathbf{1}_C(y),$$
$$Q(x, 1; dy \times \{0\}) = \psi(dy)(1 - \varepsilon \mathbf{1}_C(y)),$$
$$Q(x, 1; dy \times \{1\}) = \psi(dy)\varepsilon \mathbf{1}_C(y).$$

We let (X_n, I_n) denote the canonical process on $(\Omega, \mathcal{F}) = (\mathcal{M}^{\mathbb{N}}, \mathcal{B}(\mathcal{M})^{\otimes \mathbb{N}})$, $\mathcal{F}_n = \sigma((X_i, I_i)_{i \leq n})$, and for each $\nu \in \mathcal{P}(\mathcal{M})$, $\mathbb{P}_\nu$ the Markov measure on Ω making (X_n, I_n) a Markov chain with kernel Q (with respect to $(\mathcal{F}_n)$) and initial law ν. As usual we write $\mathbb{P}_{x,i}$ for $\mathbb{P}_{\delta_{(x,i)}}$. We shall also use the following convenient notation:

$$\mathbb{P}_x := \mathbb{P}_{x,0} \text{ if } x \in M \setminus C,$$

$$\mathbb{P}_x := (1 - \varepsilon)\mathbb{P}_{x,0} + \varepsilon \mathbb{P}_{x,1} \text{ if } x \in C.$$

Let $\mathcal{G}_n = \sigma((X_i)_{i \leq n})$. It is not hard to verify (but still a good and recommended exercise) that

$$\mathbb{P}_\nu(X_{n+1} \in \cdot | \mathcal{G}_n) = P(X_n, \cdot)$$

for all $n \geq 1$ and $\nu \in \mathcal{P}(\mathcal{M})$, and that

$$\mathbb{P}_x(X_1 \in \cdot) = P(x, \cdot).$$

This shows that $(X_n)_{n \geq 0}$ is a Markov chain with kernel P and initial value $X_0 = x$ on $(\Omega, \mathcal{F}, (\mathcal{G}_n), \mathbb{P}_x)$.

We claim that:

(a) $C \times \{1\}$ is a recurrent aperiodic atom for Q;

(b) If, for some $p \geq 1$, $\sup_{x \in C} \mathbb{E}_x(\tau_C^p) < \infty$, then there exist $a, b \geq 0$ such that for all $(x, i) \in \mathcal{M}$

$$\mathbb{E}_{x,i}(\tau_{C \times 1}^p) \leq a\mathbb{E}_x(\tau_C^p) + b;$$

(c) If, for some $\lambda_0 > 0$, $\sup_{x \in C} \mathbb{E}_x(e^{\lambda_0 \tau_C}) < \infty$, then there exist $a \geq 0$ and $0 < \lambda \leq \lambda_0$ such that for all $(x, i) \in \mathcal{M}$

$$\mathbb{E}_{x,i}(e^{\lambda \tau_{C \times 1}}) \leq a\mathbb{E}_x(e^{\lambda \tau_C}).$$

Assume the claims are proved. Then, by step 1, (X_n, I_n) is positive recurrent, and so is (X_n). As $n \to \infty$, the sequence of probability measures $P^n(x, \cdot) = \mathbb{P}_x(X_n \in \cdot)$ converges in total variation toward π, the invariant probability measure of P. If $\sup_{x \in C} \mathbb{E}_x(\tau_C^p) < \infty$ for some $p \geq 2$, then, by (b) in the claim, $\sup_{x \in C} \mathbb{E}_{x,1}(\tau_{C \times 1}^p) < \infty$. Thus, by step 1,

$$|P^n(x, A) - \pi(A)| = |\mathbb{P}_x(X_n \in A) - \pi(A)|$$

$$\leq \frac{1}{2n^{p-1}} c(1 + \mathbb{E}_x(\tau_{C \times 1}^{p-1})) \leq \frac{1}{2n^{p-1}} c(1 + a\mathbb{E}_x(\tau_C^{p-1}) + b)$$

for every $x \in M$ and $A \in \mathcal{B}(M)$. Thus, for every $\mu \in \mathcal{P}(M)$,

$$|\mu P^n - \pi| = 2 \sup_{A \in \mathcal{B}(M)} |\mu P^n(A) - \pi(A)| \leq \frac{1}{n^{p-1}} c(1 + a\mathbb{E}_\mu(\tau_C^{p-1}) + b).$$

This proves the second assertion. The proof of the third one is similar.

We now prove the claims. Clearly $C \times \{1\}$ is an atom for Q. Identify C with the subset of $\mathcal{M}$ consisting of points (x, i) such that $x \in C$. Under this identification, $C \times \{1\} \subset C$ and we rely on Proposition 7.16 to prove the claim. By the assumption that C is recurrent for P, for all $x \in M$,

$$1 = \mathbb{P}_x(\tau_C < \infty) = \begin{cases} (1 - \varepsilon)\mathbb{P}_{x,0}(\tau_C < \infty) + \varepsilon\mathbb{P}_{x,1}(\tau_C < \infty) & \text{if } x \in C, \\ \mathbb{P}_{x,0}(\tau_C < \infty) & \text{if } x \in M \setminus C. \end{cases}$$

Thus, for all $(x, i) \in \mathcal{M}$, $\mathbb{P}_{x,i}(\tau_C < \infty) = 1$, showing that C is recurrent for Q. Also,

$$\mathbb{P}_{x,i}((X_{\tau_C}, I_{\tau_C}) \in C \times \{1\}) = \varepsilon$$

because $\mathbb{P}_{x,i}((X_{\tau_C}, I_{\tau_C}) \in C \times \{1\}|\mathcal{G}_{\tau_C}) = \varepsilon$. Thus, by Proposition 7.16 (i), $C \times \{1\}$ is recurrent for Q. We now prove that it is aperiodic. For $x \in C$, $j, k \geq 1$,

$$\mathbb{P}_{x,1}(X_{j+k} \in C, I_{j+k} = 1) = \varepsilon \mathbb{P}_{x,1}(X_{j+k} \in C) \geq \varepsilon \mathbb{E}_{x,1}(\mathbf{1}_{\tau_C=j} P^k(X_{\tau_C}, C))$$

$$\geq \varepsilon \mathbb{P}_{x,1}(\tau_C = j) \inf_{x \in C} P^k(x, C).$$

Since $C \times \{1\}$ is an atom, $\mathbb{P}_{x,1}(\tau_C = j)$ does not depend on $x \in C$ and is > 0 for some $j = j_0 \geq 1$. By aperiodicity of C for P, there exists $n_0 \in \mathbb{N}$ such that for all $k \geq n_0$

$$\inf_{x \in C} P^k(x, C) > 0.$$

Therefore $\inf_{x \in C} \mathbb{P}_{x,1}(X_k \in C, I_k = 1) > 0$ for all $k \geq n_0 + j_0$. This proves aperiodicity and concludes the proof of claim (a).

If $\sup_{x \in C} \mathbb{E}_x(\tau_C^p) < \infty$ for some $p \geq 1$, then $\sup_{x \in C} \mathbb{E}_{x,i}(\tau_C^p) < \infty$ for $i \in \{0, 1\}$, and by Proposition 7.16 (ii), $\sup_{x \in C} \mathbb{E}_{x,i}(\tau_{C \times \{1\}}^p) < \infty$. Now

$$\tau_{C \times \{1\}} \leq \tau_C + \tau_{C \times \{1\}} \circ \theta_{\tau_C},$$

so that

$$\mathbb{E}_{x,i}(\tau_{C \times \{1\}}^p) \leq 2^{p-1}(\mathbb{E}_{x,i}(\tau_C^p) + \sup_{x \in C, i=0,1} \mathbb{E}_{x,i}(\tau_{C \times \{1\}}^p)) = a\mathbb{E}_{x,i}(\tau_C^p) + b.$$

Claim (c) is proved similarly. $\qquad\square$

Remark 8.19 It is interesting to compare Theorems 8.12 and 8.18 (iii). Under the assumptions of Theorem 8.12, the set $C = \{V \leq R\}$ with $R \geq \frac{2\kappa}{1-\rho}$ satisfies condition (iii) of Theorem 8.18 (with $\lambda_0 = \frac{1-\rho}{2}$). This follows from Proposition 7.12 (iii) or Proposition 7.14 (choose $\phi(s) = \lambda_0 s$). Then, by Theorem 8.18, $|P^n f(x) - \pi(f)| \leq e^{-\lambda n} c(1 + V(x))\|f\|_\infty$ for all $f \in B(M)$. Observe however that the conclusion of Theorem 8.12 is stronger, in the sense that it allows to deal with functions that are unbounded but majorized by $1 + V$ times a constant.

8.3 Convergence in Wasserstein Distance

Let H be a separable real Hilbert space with norm $\| \cdot \|$ and let P be a Markov kernel on $(H, \mathcal{B}(H))$. Let $F(H)$ be the space of bounded functions $f : H \to \mathbb{R}$ with bounded and continuous Fréchet derivative, as defined in Sect. 5.3.2. Recall from Sect. 5.3 that for a bounded metric d on H, $\mathrm{Lip}_1(d)$ denotes the set of Borel measurable functions $\phi : H \to \mathbb{R}$ such that $|\phi(x) - \phi(y)| \leq d(x, y)$ for every

$x, y \in H$. Also recall that

$$\|\mu - \nu\|_d := \sup_{\phi \in \text{Lip}_1(d)} (\mu\phi - \nu\phi), \quad \mu, \nu \in \mathcal{P}(H).$$

If (H, d) is Polish, then

$$\|\mu - \nu\|_d = W_1(\mu, \nu) := \inf_{\Gamma \in \mathcal{C}(\mu, \nu)} \int_{H^2} d(x, y) \, \Gamma(dx, dy)$$

and the metric W_1 on $\mathcal{P}(H)$ is called the Wasserstein distance of order 1 (or simply Wasserstein distance) corresponding to d, see Remark 5.36.

The following theorem provides conditions under which the mapping $\mu \mapsto \mu P$ is a strong contraction in a certain Wasserstein distance. It is a discrete-time version of Theorem 2.5 in [35], which was formulated for a continuous-time Markov semigroup.

Theorem 8.20 (Hairer, Mattingly) *Assume that there exist constants $\alpha \in (0, 1)$ and $C > 0$ such that for every $f \in F(H)$, one has $Pf \in F(H)$ and*

$$\|\nabla Pf\|_\infty \le C\|f\|_\infty + \alpha\|\nabla f\|_\infty. \tag{8.7}$$

Define

$$\gamma := \frac{1 - \alpha}{2C}, \quad B := \left\{(x, y) \in H^2 : \|x - y\| \le \gamma/2\right\},$$

and assume that

$$a := \inf_{x, y \in H} \sup \left\{\Gamma(B) : \Gamma \in \mathcal{C}(\delta_x P, \delta_y P)\right\} > 0. \tag{8.8}$$

Then there exists $\beta \in (0, 1)$ such that

$$\|\mu P - \nu P\|_d \le \beta\|\mu - \nu\|_d, \quad \forall \mu, \nu \in \mathcal{P}(H)$$

for the bounded metric $d(x, y) := 1 \wedge (\gamma^{-1}\|x - y\|)$. One can choose $\beta = \max\{(1 + \alpha)/2, 1 - \frac{a}{2}\}$.

Notice that the condition in (8.7) implies that P is asymptotically strong Feller (see Theorem 5.30). The condition in (8.8) is relatively strong but, as we shall see, allows for a short and transparent proof. In [35], Hairer and Mattingly also formulate a set of Lyapunov-type conditions which imply that $\mu \mapsto \mu P$ is a strong contraction in the Wasserstein distance corresponding to the metric

$$d(x, y) := \inf_{\gamma} \int_0^1 V(\gamma(s)) \|\dot{\gamma}(s)\| \, ds,$$

where V is a suitable Lyapunov function and the infimum is taken over absolutely continuous paths $\gamma : [0, 1] \to H$ such that $\gamma(0) = x$ and $\gamma(1) = y$. The latter result is more broadly applicable, in particular to the two-dimensional stochastic Navier–Stokes equation.

Proof *(Theorem 8.20)* We first show that there exists $\beta \in (0, 1)$ such that

$$\|\delta_x P - \delta_y P\|_d \leq \beta d(x, y), \quad \forall x, y \in H. \tag{8.9}$$

Let $\phi \in \mathrm{Lip}_1(d)$. By Remark 5.30, there exists a sequence $(\phi_n)_{n \geq 1}$ in $F(H) \cap \mathrm{Lip}_1(d)$ such that

$$\lim_{n \to \infty} \phi_n(x) = \phi(x), \quad \forall x \in H.$$

Define $\tilde{\phi}$ and $(\tilde{\phi}_n)_{n \geq 1}$ as in the proof of Theorem 5.29. Then $\tilde{\phi}_n \in F(H) \cap \mathrm{Lip}_1(d)$ and $\|\nabla \tilde{\phi}_n\|_\infty \leq \gamma^{-1}$ for every $n \in \mathbb{N}^*$. By assumption, for every $n \in \mathbb{N}^*$, one has $P\tilde{\phi}_n \in F(H)$ and

$$\|\nabla P\tilde{\phi}_n\|_\infty \leq C\|\tilde{\phi}_n\|_\infty + \alpha\|\nabla \tilde{\phi}_n\|_\infty \leq C + \frac{2\alpha C}{1 - \alpha}.$$

As shown in the proof of Theorem 5.29, for $n \in \mathbb{N}^*$ and $x, y \in H$,

$$P\phi_n(x) - P\phi_n(y) \leq \|x - y\|\|\nabla P\tilde{\phi}_n\|_\infty \leq \|x - y\|\frac{C(1 + \alpha)}{1 - \alpha}.$$

Let $x, y \in H$ such that $\|x - y\| < \gamma$. Then $d(x, y) = \gamma^{-1}\|x - y\|$, so

$$P\phi_n(x) - P\phi_n(y) \leq d(x, y)\gamma\frac{C(1 + \alpha)}{1 - \alpha} = d(x, y)(1 + \alpha)/2.$$

By bounded convergence,

$$P\phi(x) - P\phi(y) \leq d(x, y)(1 + \alpha)/2.$$

As this estimate holds for all $\phi \in \mathrm{Lip}_1(d)$,

$$\|\delta_x P - \delta_y P\|_d \leq d(x, y)(1 + \alpha)/2.$$

Now let $x, y \in H$ such that $\|x - y\| \geq \gamma$. Then there exists $\tilde{\Gamma} \in \mathcal{C}(\delta_x P, \delta_y P)$ such that

$$\tilde{\Gamma}(B) \geq a/2.$$

The space H with the norm-induced metric is Polish, and d is a bounded continuous metric on H. By Kantorovich–Rubinstein duality (Theorem 5.34),

$$\|\delta_x P - \delta_y P\|_d = \inf_{\Gamma \in \mathcal{C}(\delta_x P, \delta_y P)} \int_{H^2} d(a, b)\, \Gamma(da, db)$$

$$\leq \int_{H^2} d(a, b)\, \tilde{\Gamma}(da, db)$$

$$= \int_B d(a, b)\, \tilde{\Gamma}(da, db) + \int_{H^2 \setminus B} d(a, b)\, \tilde{\Gamma}(da, db).$$

For $(a, b) \in B$, one has

$$d(a, b) \leq \gamma^{-1}\|a - b\| \leq 1/2.$$

Hence

$$\int_B d(a, b)\, \tilde{\Gamma}(da, db) \leq \frac{1}{2}\tilde{\Gamma}(B).$$

And since the metric d is bounded by 1,

$$\int_{H^2 \setminus B} d(a, b)\, \tilde{\Gamma}(da, db) \leq \tilde{\Gamma}(H^2 \setminus B) = 1 - \tilde{\Gamma}(B).$$

As a result,

$$\|\delta_x P - \delta_y P\|_d \leq 1 - \frac{1}{2}\tilde{\Gamma}(B) \leq 1 - \frac{a}{2}.$$

Our assumption that $\|x - y\| \geq \gamma$ implies that $d(x, y) = 1$. Hence

$$\|\delta_x P - \delta_y P\|_d \leq d(x, y)\left(1 - \frac{a}{2}\right).$$

This proves (8.9) for $\beta = \max\{(1 + \alpha)/2; 1 - \frac{a}{2}\}$. To complete the proof of Theorem 8.20, let $\mu, \nu \in \mathcal{P}(H)$. By Theorem 5.34, there exists $\Gamma^* \in \mathcal{C}(\mu, \nu)$ such that

$$\|\mu - \nu\|_d = \int_{H^2} d(x, y)\, \Gamma^*(dx, dy).$$

Let $\phi \in \mathrm{Lip}_1(d)$. Then, by Exercise 5.38,

$$(\mu P)\phi - (\nu P)\phi = \int_{H^2} ((\delta_x P)\phi - (\delta_y P)\phi)\, \Gamma^*(dx, dy). \qquad (8.10)$$

For $x, y \in H$,

$$(\delta_x P)\phi - (\delta_y P)\phi \le \|\delta_x P - \delta_y P\|_d \le \beta d(x, y).$$

Hence, the right-hand side of (8.10) is dominated by

$$\beta \int_{H^2} d(x, y)\, \Gamma^*(dx, dy) = \beta \|\mu - \nu\|_d.$$

Taking the supremum for the left-hand side of (8.10) over all $\phi \in \mathrm{Lip}_1(d)$ yields the desired contraction estimate. $\qquad\square$

Corollary 8.21 *Under the assumptions of Theorem 8.20, the Markov kernel P admits a unique invariant probability measure π and there exists $\beta \in (0, 1)$ such that for every $\mu \in \mathcal{P}(H)$,*

$$\|\mu P^n - \pi\|_d \le \beta^n \|\mu - \pi\|_d, \quad \forall n \in \mathbb{N}^*.$$

Proof Clearly, d induces the same topology on H as the metric induced by $\|\cdot\|$. Then (H, d) is a Polish space with a bounded metric. By Remark 5.36, $\mathcal{P}(H)$ endowed with the metric $(\mu, \nu) \mapsto \|\mu - \nu\|_d$ is Polish. Since $\mu \mapsto \mu P$ is a strong contraction on this complete metric space, the Banach fixed point theorem yields existence and uniqueness of the invariant probability measure π. For $\mu \in \mathcal{P}(H)$ and $n \in \mathbb{N}^*$, one has

$$\|\mu P^n - \pi\|_d = \|\mu P^n - \pi P^n\|_d \le \beta^n \|\mu - \pi\|_d,$$

where β is the constant from Theorem 8.20. $\qquad\square$

Appendix A
Monotone Class and Martingales

A.1 Monotone Class Theorem

A set $H \subset B(M)$ is said to be *stable by bounded monotone convergence* if $f_n \in H$ and $0 \le f_n \le f_{n+1} \le 1$ implies that $f = \lim_n f_n \in H$.

Theorem A.1 (Monotone Class Theorem) *Let* $H \subset B(M)$ *be a vector space of bounded functions containing the constant functions and stable by bounded monotone convergence. Let* $C \subset B(M)$ *be a set stable by multiplication and let* $\sigma(C)$ *denote the σ-field generated by C (i.e., the smallest σ-field for which the elements of C are measurable). If $C \subset H$, then H contains every bounded $\sigma(C)$-measurable function.*

A.2 Conditional Expectation

We recall here the definition of conditional expectation and give some of its basic properties. More details and proofs can be found in standard textbooks such as [7].

Let $(\Omega, \mathcal{F}, P)$ be a probability space and let $\mathcal{B}$ be a σ-field contained in $\mathcal{F}$. Let X be a real-valued random variable such that $E(|X|) < \infty$. Then there exists a real-valued random variable Z with $E(|Z|) < \infty$ such that

(i) Z is $\mathcal{B}-$measurable;
(ii) For all $A \in \mathcal{B}$, we have

$$E(Z\mathbf{1}_A) = E(X\mathbf{1}_A).$$

© The Author(s), under exclusive license to Springer Nature Switzerland AG 2022
M. Benaïm, T. Hurth, *Markov Chains on Metric Spaces*, Universitext,
https://doi.org/10.1007/978-3-031-11822-7

The random variable Z is unique in the following sense: If Z' is any other random variable satisfying $E(|Z'|) < \infty$ and the conditions in (i) and (ii), then $P(Z' = Z) = 1$. In other words, the space of equivalence classes $L^1(\Omega, \mathcal{B}, P)$ has a unique element Z satisfying the condition in (ii). This element of $L^1(\Omega, \mathcal{B}, P)$ is called the *conditional expectation* of X given $\mathcal{B}$, and is denoted by $E(X|\mathcal{B})$. If we write $Y = E(X|\mathcal{B})$ for some $\mathcal{B}$-measurable random variable Y, we mean that Y is a representative of the equivalence class $E(X|\mathcal{B})$.

One can also define conditional expectation for nonnegative random variables: Let $X : \Omega \to [0, \infty]$ be measurable, i.e., $\{\omega \in \Omega : X(\omega) \in A\} \in \mathcal{F}$ for every set $A \subset [0, \infty]$ such that $A \setminus \{\infty\}$ is a Borel subset of $[0, \infty)$. For every $n \in \mathbb{N}$, let $X_n := X \wedge n$ and let Z_n be a $\mathcal{B}$-measurable random variable such that $E(|Z_n|) < \infty$ and $E(Z_n 1_A) = E(X_n 1_A)$ for every $A \in \mathcal{B}$. By changing the values of (Z_n) on a set of measure 0 if necessary, one can assume that $(Z_n(\omega))_{n \in \mathbb{N}}$ is nondecreasing for every $\omega \in \Omega$. The function

$$Z(\omega) := \lim_{n \to \infty} Z_n(\omega)$$

then maps from Ω to $[0, \infty]$ and satisfies the conditions in (i) and (ii). If $Z' : \Omega \to [0, \infty]$ is any other random variable satisfying (i) and (ii), then $P(Z = Z') = 1$. On the set of $\mathcal{B}$-measurable functions from Ω to $[0, \infty]$, consider the equivalence relation given by equality P-almost surely. The conditional expectation of X given $\mathcal{B}$, denoted by $E(X|\mathcal{B})$, is defined as the unique equivalence class that satisfies (ii).

Theorem A.2 (Properties of Conditional Expectation) *Let X be a random variable, with $E(|X|) < \infty$ or $X \in [0, \infty]$, and let $\mathcal{B}$ be a σ-field contained in $\mathcal{F}$. Then*

(i) $E(E(X|\mathcal{B})) = E(X)$;
(ii) *If $E(|X|) < \infty$ (resp. $X \in [0, \infty]$), we have for every $\mathcal{B}$-measurable random variable Y with $E(|XY|) < \infty$ (resp. $Y \in [0, \infty]$)*

$$E(XY|\mathcal{B}) = YE(X|\mathcal{B}),$$

with the convention that $0 \cdot \infty = 0$;
(iii) *For every σ-field $\mathcal{A}$ contained in $\mathcal{B}$, we have*

$$E(E(X|\mathcal{B})|\mathcal{A}) = E(X|\mathcal{A}).$$

This is often called tower property.

A.3   Martingales

Here, we recall the few results from martingale theory that are used in this course. As for conditional expectation, there are many introductory texts on probability theory that provide more details and proofs, e.g., [69] or [7].

Let $(\Omega, \mathcal{F}, \mathbb{F}, \mathsf{P})$ be a filtered probability space. We let $\mathcal{F}_\infty$ denote the σ-field generated by $\cup_{n \geq 0} \mathcal{F}_n$. A sequence (M_n) of adapted (i.e., M_n is $\mathcal{F}_n$-measurable) and L^1 real-valued random variables is called a *martingale* (respectively a *submartingale*, respectively a *supermartingale*) if

$$\mathsf{E}(M_{n+1}|\mathcal{F}_n) = M_n \text{ resp. } \geq, \text{ resp. } \leq$$

for all $n \geq 0$.

A simple, but useful consequence of Jensen's inequality is the following result.

Proposition A.3 *Let (M_n) be a martingale (resp. a submartingale) and ϕ a convex function (resp. a convex nondecreasing function) such that $\phi(M_n) \in L^1$. Then $(\phi(M_n))$ is a submartingale.*

It is often useful to extend the martingale (submartingale, supermartingale) property to stopping times. Doob's optional stopping theorem shows that this can be done for bounded stopping times.

Theorem A.4 (Optional Stopping)   *Let $M = (M_n)$ be a martingale (resp. submartingale, supermartingale).*

 (i) *If T is a stopping time, then $(M_{n \wedge T})_{n \geq 0}$ is a martingale (resp. submartingale, supermartingale);*

(ii) *If $S \leq T$ are stopping times bounded by some constant N, then*

$$E(M_T|\mathcal{F}_S) = M_S, \text{ resp. } \geq, \text{ resp. } \leq.$$

Proof

 (*i*)   For all $n \in \mathbb{N}$

$$M_{(n+1) \wedge T} - M_{n \wedge T} = (M_{n+1} - M_n)\mathbf{1}_{\{T > n\}}.$$

Taking the conditional expectation with respect to $\mathcal{F}_n$ proves the result.

(ii) Assume (M_n) is a martingale. Proving that $E(M_T|\mathcal{F}_S) = M_S$ amounts to proving that for all $A \in \mathcal{F}_S$ and $0 \le k \le N$, $E(M_T 1_{A \cap \{S=k\}}) = E(M_k 1_{A \cap \{S=k\}})$. One has

$$E(M_T 1_{A \cap \{S=k\}}) = \sum_{i=k}^{N} E(M_i 1_{\{T=i\}} 1_{A \cap \{S=k\}}) = \sum_{i=k}^{N} E(E(M_N|\mathcal{F}_i) 1_{\{T=i\}} 1_{A \cap \{S=k\}})$$

$$= \sum_{i=k}^{N} E(E(M_N 1_{\{T=i\}} 1_{A \cap \{S=k\}}|\mathcal{F}_i)) = E(M_N 1_{A \cap \{S=k\}})$$

$$= E(E(M_N 1_{A \cap \{S=k\}}|\mathcal{F}_k)) = E(M_k 1_{A \cap \{S=k\}}).$$

The proof for sub- and supermartingales is similar.

□

Corollary A.5 (Doob's Inequality) *Let (X_n) be a nonnegative submartingale. Then, for all $\alpha > 0$,*

$$P(\sup_{0 \le i \le N} X_i \ge \alpha) \le \frac{E(X_N)}{\alpha}.$$

Proof Let $T = \min\{i \ge 0 : X_i \ge \alpha\}$. Then $T \wedge N$ is a stopping time bounded by N so that, by the optional stopping theorem,

$$E(X_N) \ge E(X_{N \wedge T}) = E(X_N 1_{T>N}) + E(X_T 1_{T \le N}) \ge \alpha P(T \le N).$$

□

The two following theorems are classical convergence results due to Doob.

Theorem A.6 *Let (M_n) be a submartingale. Assume that $\sup_n E(M_n^+) < \infty$. Then there exists $M_\infty \in L^1$ such that $M_n \to M_\infty$ almost surely.*

Theorem A.7 *Let (M_n) be a martingale. Then the following assertions are equivalent:*

(a) *(M_n) is uniformly integrable;*
(b) *(M_n) converges almost surely and in L^1 to some random variable M_∞;*
(c) *$M_n = E(M|\mathcal{F}_n)$ for some $M \in L^1$.*

Furthermore, in case (c), $\lim_{n \to \infty} M_n = M_\infty = E(M|\mathcal{F}_\infty)$.

Let (M_n) be an L^2-martingale (i.e., $M_n \in L^2$). The *predictable quadratic variation* of (M_n) is the process $(\langle M \rangle_n)$ recursively defined as

$$\langle M \rangle_0 = 0, \langle M \rangle_{n+1} - \langle M \rangle_n = \mathsf{E}((M_{n+1} - M_n)^2 | \mathcal{F}_n) = \mathsf{E}(M_{n+1}^2 | \mathcal{F}_n) - M_n^2.$$

Note that $(\langle M \rangle_n)$ is nondecreasing, predictable (i.e., M_n is $\mathcal{F}_{n-1}$-measurable) and that $(M_n^2 - \langle M \rangle_n)_n$ is a zero-mean martingale. We let $\langle M \rangle_\infty = \lim_{n \to \infty} \langle M \rangle_n$.

Theorem A.8 (Strong Law of Large Numbers) *Let (M_n) be an L^2-martingale. Then*

(i) *If $\mathsf{E}(\langle M \rangle_\infty) = \sum_{k \geq 0} \mathsf{E}((M_{k+1} - M_k)^2) < \infty$, then (M_n) converges almost surely and in L^2 to some random variable M_∞;*

(ii) *On $\langle M_\infty \rangle < \infty$, (M_n) converges almost surely to some finite random variable M_∞;*

(iii) *On $\langle M \rangle_\infty = \infty$, $\lim_{n \to \infty} \frac{M_n}{\langle M \rangle_n} = 0$ a.s.*

(iv) *If $\sup_n \mathsf{E}(\frac{\langle M \rangle_n}{n}) < \infty$, then $\lim_{n \to \infty} \frac{M_n}{n} = 0$ a.s.*

Proof We only prove the last statement, which allows for a short proof and which is all that is needed in this book. By Doob's inequality, for all $n \in \mathbb{N}$,

$$\mathsf{P}\left(\sup_{2^n \leq k \leq 2^{n+1}} \frac{|M_k|}{k} \geq \varepsilon \right) \leq \mathsf{P}(\sup_{k \leq 2^{n+1}} |M_k|^2 \geq \varepsilon^2 2^{2n}) \leq \frac{1}{\varepsilon^2 2^{2n}} \mathsf{E}(\langle M \rangle_{2^n}) \leq C \frac{1}{\varepsilon^2 2^n}$$

and the result follows from the Borel-Cantelli lemma.   □

Bibliography

1. Aldous, D., Fill, J.: Reversible Markov chains and random walks on graphs. https://www.stat.berkeley.edu/users/aldous/RWG/book.html (1995)
2. Arnold, L., Kliemann, W.: On unique ergodicity for degenerate diffusions. Stochastics **21**(1), 41–61 (1987). MR 899954
3. Aurzada, F., Döring, H., Ortgiese, M., Scheutzow, M.: Moments of recurrence times for Markov chains. Electron. Commun. Probab. **16**, 296–303 (2011)
4. Bakhtin, Y.: Some topics in ergodic theory. https://cims.nyu.edu/%7Ebakhtin/2015-ergodic.pdf (2015)
5. Bakhtin, Y., Hurth, T.: Invariant densities for dynamical systems with random switching. Nonlinearity **25**(10), 2937–2952 (2012). MR 2979976
6. Benaim, M.: Stochastic persistence. Preprint. arXiv:1806.08450 (2018)
7. Benaim, M., El Karoui, N.: Promenade aléatoire : Chaînes de Markov et Simulation, Editions de l'Ecole Polytechnique (2005)
8. Benaïm, M., Le Borgne, S., Malrieu, F., Zitt, P.-A.: Qualitative properties of certain piecewise deterministic Markov processes. Ann. Inst. Henri Poincaré Probab. Stat. **51**(3), 1040–1075 (2015)
9. Benaïm, M., Lobry, C.: Lotka-Volterra with randomly fluctuating environments or "How switching between beneficial environments can make survival harder". Ann. Appl. Probab. **26**(6), 3754–3785 (2016)
10. Benaïm, M., Champagnat, N., Oçafrain, W., Villemonais, D.: Degenerate processes killed at the boundary of a domain. Preprint. arXiv:2103.08534 (2021)
11. Billingsley, P.: Convergence of Probability Measures, 2nd ed., Wiley Series in Probability and Statistics: Probability and Statistics. John Wiley & Sons, Inc., New York (1999). A Wiley-Interscience Publication
12. Blumenthal, R.M., Corson, H.H.: On continuous collections of measures. In: Proceedings of the Sixth Berkeley Symposium on Mathematical Statistics and Probability (Univ. California, Berkeley, Calif., 1970/1971), Vol. II: Probability theory, pp. 33–40 (1972)
13. Bony, J.M.: Principe du maximum, inégalite de Harnack et unicité du problème de Cauchy pour les opérateurs elliptiques dégénérés. Ann. Inst. Fourier (Grenoble) **19**(fasc. 1), 277–304 xii (1969). MR 262881
14. Cerrai, S.: A Hille–Yosida theorem for weakly continuous semigroups. In: Semigroup Forum, vol. 49, pp. 349–367. Springer (1994)
15. Chow, W.L.: Über Systeme von linearen partiellen Differentialgleichungen erster Ordnung. Math. Ann. **117**, 98–105 (1939). MR 1880

© The Author(s), under exclusive license to Springer Nature Switzerland AG 2022
M. Benaïm, T. Hurth, *Markov Chains on Metric Spaces*, Universitext,
https://doi.org/10.1007/978-3-031-11822-7

16. Chung, K.L.: Contributions to the theory of Markov chains. II. Trans. Amer. Math. Soc. **76**, 397–419 (1954)
17. Cornfeld, I.P., Fomin, S.V., Sinai, Ya.G.: Ergodic Theory, Grundlehren der Mathematischen Wissenschaften [Fundamental Principles of Mathematical Sciences], vol. 245. Springer, New York (1982). Translated from the Russian by A. B. Sosinskiĭ
18. Dellacherie, C., Meyer, P.-A.: Probabilités et potentiel. Chapitres IX à XI, revised ed., Publications de l'Institut de Mathématiques de l'Université de Strasbourg [Publications of the Mathematical Institute of the University of Strasbourg], XVIII. Hermann, Paris (1983), Théorie discrète du potentiel. [Discrete potential theory], Actualités Scientifiques et Industrielles [Current Scientific and Industrial Topics], 1410
19. Diaconis, P., Freedman, D.: Iterated random functions. SIAM Rev. **41**(1), 45–76 (1999)
20. Douc, R., Moulines, E., Priouret, P., Soulier, P.: Markov Chains. Springer Series in Operations Research and Financial Engineering. Springer, Cham (2018). MR 3889011
21. Dudley, R.M.: Real Analysis and Probability, Cambridge Studies in Advanced Mathematics, vol. 74. Cambridge University Press, Cambridge (2002). Revised reprint of the 1989 original
22. Duflo, M.: Random Iterative Models, Applications of Mathematics (New York), vol. 34. Springer, Berlin (1997), Translated from the 1990 French original by Stephen S. Wilson and revised by the author
23. Dugundji, J.: Topology. Allyn and Bacon, Boston (1966)
24. Einsiedler, M., Ward, T.: Ergodic Theory with a View Towards Number Theory, Graduate Texts in Mathematics, vol. 259. Springer-Verlag London, London (2011)
25. Erdös, P.: On a family of symmetric Bernoulli convolutions. Amer. J. Math. **61**, 974–976 (1939). MR 311
26. Ethier, S.N., Kurtz, T.G.: Markov Processes: Characterization and Convergence. Wiley Series in Probability and Statistics (1986)
27. Folland, G.B.: Real Analysis, 2nd edn., Pure and Applied Mathematics (New York). John Wiley & Sons, New York (1999). Modern techniques and their applications, A Wiley-Interscience Publication
28. Furman, A.: A brief introduction to ergodic theory, Unpublished. http://homepages.math.uic.edu/~furman/papers.html (2012)
29. Furstenberg, H.: Strict ergodicity and transformation of the torus. Amer. J. Math. **83**, 573–601 (1961)
30. Gilbarg, D., Trudinger, N.S.: Elliptic Partial Differential Equations of Second Order, Classics in Mathematics. Springer, Berlin (2001). Reprint of the 1998 edition.
31. Hairer, M.: Ergodic properties of Markov processes, Lecture given at the University of Warwick (2006)
32. Hairer, M.: Ergodic properties of a class of non-Markovian processes. In: Trends in Stochastic Analysis, London Math. Soc. Lecture Note Ser., vol. 353, pp. 65–98. Cambridge Univ. Press, Cambridge (2009)
33. Hairer, M.: Convergence of Markov processes, http://hairer.org/notes/Convergence.pdf (2016)
34. Hairer, M., Mattingly, J.C.: Ergodicity of the 2D Navier-Stokes equations with degenerate stochastic forcing. Ann. Math. (2) **164**(3), 993–1032 (2006)
35. Hairer, M., Mattingly, J.C.: Spectral gaps in Wasserstein distances and the 2D stochastic Navier–Stokes equations. Ann. Probab. **36**(6), 2050–2091 (2008)
36. Hairer, M., Mattingly, J.C.: Yet another look at Harris' ergodic theorem for Markov chains. In: Seminar on Stochastic Analysis, Random Fields and Applications VI, Progr. Probab., vol. 63, Birkhäuser/Springer Basel AG, Basel (2011), pp. 109–117
37. Harris, T.E.: The existence of stationary measures for certain Markov processes. In: Proceedings of the Third Berkeley Symposium on Mathematical Statistics and Probability, 1954–1955, vol. II, pp. 113–124. University of California Press, Berkeley (1956). MR 0084889
38. Hörmander, L.: Hypoelliptic second order differential equations. Acta Math. **119**, 147–171 (1967). MR 222474
39. Ichihara, K., Kunita, H.: A classification of the second order degenerate elliptic operators and its probabilistic characterization. Z. Wahrsch. Verw. Gebiete **30**(3), 235–254 (1974)

40. Ikeda, N., Watanabe, S: Stochastic Differential Equations and Diffusion Processes, North-Holland Mathematical Library, vol. 24. North-Holland Publishing Co./Kodansha, Ltd., Amsterdam/Tokyo (1981). MR 637061
41. Jurdjevic, V.: Geometric Control Theory, Cambridge Studies in Advanced Mathematics, vol. 52. Cambridge University Press, Cambridge (1997). MR 1425878
42. Katok, A., Hasselblatt, B.: Introduction to the Modern Theory of Dynamical Systems, Encyclopedia of Mathematics and its Applications, vol. 54. Cambridge University Press, Cambridge (1995), With a supplementary chapter by Katok and Leonardo Mendoza
43. Kifer, Y.: Ergodic Theory of Random Transformations, Progress in Probability and Statistics, vol. 10. Birkhäuser Boston, Boston (1986)
44. Kunita, H: Stochastic Flows and Stochastic Differential Equations, Cambridge Studies in Advanced Mathematics, vol. 24. Cambridge University Press, Cambridge (1990). MR 1070361
45. Le Gall, J.-F., et al.: Brownian Motion, Martingales, and Stochastic Calculus, vol. 274. Springer (2016)
46. Levin, D.A., Peres, Y., Wilmer, E.L.: Markov Chains and Mixing Times. American Mathematical Society, Providence (2009). With a chapter by James G. Propp and David B. Wilson
47. Lindvall, T.: Lectures on the Coupling Method, Wiley Series in Probability and Mathematical Statistics: Probability and Mathematical Statistics. Wiley, New York (1992). A Wiley-Interscience Publication
48. Mañé, R.: Ergodic Theory and Differentiable Dynamics, Ergebnisse der Mathematik und ihrer Grenzgebiete (3) [Results in Mathematics and Related Areas (3)], vol. 8. Springer, Berlin (1987). Translated from the Portuguese by Silvio Levy
49. Meyn, S., Tweedie, R.L.: Markov Chains and Stochastic Stability, 2nd edn. Cambridge University Press, Cambridge (2009). With a prologue by Peter W. Glynn
50. Moser, J.: On the volume elements on a manifold. Trans. Amer. Math. Soc. **120**, 286–294 (1965)
51. Nualart, D.: The Malliavin Calculus and Related Topics, 2nd edn. Probability and its Applications (New York). Springer, Berlin (2006). MR 2200233 (2006j:60004)
52. Nummelin, E.: General Irreducible Markov Chains and Nonnegative Operators, Cambridge Tracts in Mathematics, vol. 83. Cambridge University Press, Cambridge (1984)
53. Peres, Y., Schlag, W., Solomyak, B.: Sixty years of Bernoulli convolutions. In: Fractal Geometry and Stochastics, II (Greifswald/Koserow, 1998), Progr. Probab., vol. 46, pp. 39–65 (Birkhäuser, Basel, 2000). MR 1785620
54. Peres, Y., Solomyak, B.: Absolute continuity of Bernoulli convolutions, a simple proof. Math. Res. Lett. **3**(2), 231–239 (1996)
55. Pitman, J.W.: Uniform rates of convergence for Markov chain transition probabilities. Z. Wahrsch. Verw. Gebiete **29**, 193–227 (1974)
56. Da Prato, G., Zabczyk, J.: Smoothing properties of transition semigroups in Hilbert spaces. Stoch. Stoch. Rep. **35**(2), 63–77 (1991)
57. Propp, J.G., Wilson, D.B.: Exact sampling with coupled Markov chains and applications to statistical mechanics. In: Proceedings of the Seventh International Conference on Random Structures and Algorithms (Atlanta, GA, 1995), vol. 9, pp. 223–252 (1996). MR 1611693
58. Quas, A.N.: On representations of Markov chains by random smooth maps. Bull. Lond. Math. Soc. **23**(5), 487–492 (1991)
59. Revuz, D., Yor, M.: Continuous Martingales and Brownian Motion, 3rd edn., Grundlehren der mathematischen Wissenschaften [Fundamental Principles of Mathematical Sciences], vol. 293. Springer, Berlin (1999). MR 1725357
60. Robert, P.: Stochastic Networks and Queues, French ed., Applications of Mathematics (New York), vol. 52. Springer, Berlin (2003). Stochastic Modelling and Applied Probability
61. Rudin, W.: Functional Analysis, 2nd edn., International Series in Pure and Applied Mathematics. McGraw-Hill, New York (1991). MR 1157815
62. Saloff-Coste, L.: Lectures on Finite Markov Chains, Lectures on Probability Theory and Statistics (Saint-Flour, 1996), Lecture Notes in Math., vol. 1665, pp. 301–413. Springer, Berlin (1997)

63. Shiryaev, A.N.: Probability, 2nd edn., Graduate Texts in Mathematics, vol. 95. Springer, New York (1996). Translated from the first (1980) Russian edition by R. P. Boas
64. Solomyak, B.: On the random series $\sum \pm \lambda^n$ (an Erdös problem). Ann. Math. (2) **142**(3), 611–625 (1995)
65. Stroock, D.W., Varadhan, S.R.S.: On the support of diffusion processes with applications to the strong maximum principle. In: Proceedings of the Sixth Berkeley Symposium on Mathematical Statistics and Probability (Univ. California, Berkeley, Calif., 1970/1971), Vol. III: Probability Theory, pp. 333–359 (1972). MR 0400425
66. Sussmann, H.J., Jurdjevic, V.: Controllability of nonlinear systems. J. Differential Equations **12**, 95–116 (1972). MR 0338882 (49 #3646)
67. Varjú, P.P.: Recent progress on Bernoulli convolutions. Preprint. arXiv:1608.04210 (2016)
68. Villani, C.: Optimal Transport, Grundlehren der Mathematischen Wissenschaften [Fundamental Principles of Mathematical Sciences], vol. 338. Springer, Berlin (2009). Old and new
69. Williams, D.: Probability with Martingales, Cambridge Mathematical Textbooks. Cambridge University Press, Cambridge (1991)
70. Woess, W.: Random Walks on Infinite Graphs and Groups, Cambridge Tracts in Mathematics, vol. 138. Cambridge University Press, Cambridge (2000)

List of Symbols

$(\nu_n)_{n \geq 1}$, 56
$(\theta_n)_{n \geq 1}$, 37
(F, m) , 38
$[G, H]$, 121
$\alpha \ll \beta$, 59
$\alpha \perp \beta$, 59
$\delta(\cdot, \cdot)$, 92
δ_x , 2
$\langle M \rangle_n$, 185
$\mathsf{Int}(C)$, 156
μf , xv
μP , 1
$\mu_n \Rightarrow \mu$, 47
∂A , 48
$\subset$, 2
τ_C , 20
τ_x , 11
$\mathsf{Lip}_1(d)$, 92
Γ , 87
Γ_C , 87
Γ_x , 87
Ω , 2
$\Phi^* \rho$, 40
Θ^n , 6
$|\cdot|_d$, 92
$|\cdot|_{bl}$, 50
$\{P_t\}_{t \geq 0}$, 8
$A \triangle B$, 67
$a \wedge b$, 96

© The Author(s), under exclusive license to Springer Nature Switzerland AG 2022   191
M. Benaïm, T. Hurth, *Markov Chains on Metric Spaces*, Universitext,
https://doi.org/10.1007/978-3-031-11822-7

Index

Printed in the United States
by Baker & Taylor Publisher Services